Lecture Notes in Electrical Engineering

Volume 493

**** Indexing: The books of this series are submitted to ISI Proceedings, EI-Compendex, SCOPUS, MetaPress, Springerlink ****

Lecture Notes in Electrical Engineering (LNEE) is a book series which reports the latest research and developments in Electrical Engineering, namely:

- Communication, Networks, and Information Theory
- Computer Engineering
- Signal, Image, Speech and Information Processing
- Circuits and Systems
- Bioengineering
- Engineering

The audience for the books in LNEE consists of advanced level students, researchers, and industry professionals working at the forefront of their fields. Much like Springer's other Lecture Notes series, LNEE will be distributed through Springer's print and electronic publishing channels.

For general information about this series, comments or suggestions, please use the contact address under "service for this series".

To submit a proposal or request further information, please contact the appropriate Springer Publishing Editors:

Asia:

China, *Jessie Guo, Assistant Editor* (jessie.guo@springer.com) (Engineering)

India, *Swati Meherishi, Senior Editor* (swati.meherishi@springer.com) (Engineering)

Japan, *Takeyuki Yonezawa, Editorial Director* (takeyuki.yonezawa@springer.com) (Physical Sciences & Engineering)

South Korea, *Smith (Ahram) Chae, Associate Editor* (smith.chae@springer.com) (Physical Sciences & Engineering)

Southeast Asia, *Ramesh Premnath, Editor* (ramesh.premnath@springer.com) (Electrical Engineering)

South Asia, *Aninda Bose, Editor* (aninda.bose@springer.com) (Electrical Engineering)

Europe:

Leontina Di Cecco, Editor (Leontina.dicecco@springer.com)

(Applied Sciences and Engineering; Bio-Inspired Robotics, Medical Robotics, Bioengineering; Computational Methods & Models in Science, Medicine and Technology; Soft Computing; Philosophy of Modern Science and Technologies; Mechanical Engineering; Ocean and Naval Engineering; Water Management & Technology)

(christoph.baumann@springer.com)

(Heat and Mass Transfer, Signal Processing and Telecommunications, and Solid and Fluid Mechanics, and Engineering Materials)

North America:

Michael Luby, Editor (michael.luby@springer.com) (Mechanics; Materials)

More information about this series at http://www.springer.com/series/7818

Adamu Murtala Zungeru
S. Subashini · P. Vetrivelan
Editors

Wireless Communication Networks and Internet of Things

Select Proceedings of ICNETS2, Volume VI

Springer

Editors
Adamu Murtala Zungeru
Department of Electrical, Computer and Telecommunication Engineering
Botswana International University of Science and Technology
Palapye
Botswana

P. Vetrivelan
School of Electronics Engineering
VIT University, Chennai
Chennai, Tamil Nadu
India

S. Subashini
School of Electronics Engineering
VIT University, Chennai
Chennai, Tamil Nadu
India

ISSN 1876-1100 ISSN 1876-1119 (electronic)
Lecture Notes in Electrical Engineering
ISBN 978-981-10-8662-5 ISBN 978-981-10-8663-2 (eBook)
https://doi.org/10.1007/978-981-10-8663-2

Library of Congress Control Number: 2018933495

Printed on acid-free paper

This Springer imprint is published by the registered company Springer Nature Singapore Pte Ltd.
part of Springer Nature
The registered company address is: 152 Beach Road, #21-01/04 Gateway East, Singapore 189721, Singapore

Preface

The International Symposium-F on "**Wireless Communication Networking and IoT**" was successfully held at VIT University, Chennai, during March 23–25, 2017, as part of ICNETS2.

The symposium convenes the researchers and practitioners working on Wireless Communication Networking and Internet of Things to discuss the current trends, breakthroughs, technical challenges, and future services and applications. The symposium featured two keynote speakers from universities abroad. The lecture by Dr. Adamu, on "Terrestrial to underground wireless sensor networks: Design factors and transmission cost" created a great inquisitiveness among the participants to explore the issues arising in wireless networks. Dr. Elizabeth Chang, University of New South Wales and Australian Defence Force Academy, Canberra, Australia, delivered a lecture on the experimental analysis in the context of "Cyber Situational Awareness for CPS, 5G and IoT" which created an impact on creating security frameworks based on ontology.

Over the past decades, the emergence of the advanced wireless technologies has attracted many researchers and academicians to explore in the fields of next-generation networks and Internet of things. Recent advances in computing have created a new dimension of research in the field of sensor networks in connecting the physical world to the digital world. The sessions focussed on resolving the challenges in the existing communication protocols connecting the different embedded devices in different scenarios, which deserves our special attention.

In response to the call for papers, 75 manuscripts are received, out of which 26 papers are accepted. We intended for maintaining the high standards of the conference. All the papers were rigorously peer-reviewed by three expert members, and the proceedings comprises of a selection of 26 papers presented at the symposium. The proceedings will provide you an insight into selection of appropriate wireless technologies and IoT for different applications.

Palapye, Botswana Adamu Murtala Zungeru
Chennai, India S. Subashini
Chennai, India P. Vetrivelan

Acknowledgements

We would like to warmly thank scientific committee members, program committee members, as well as the external reviewers, for their efforts as well as their valuable inputs in the review process in selecting the best papers. We would also like to acknowledge the speakers and the participants who came all the way to present their work. We also thank all the professors who agreed to be the chair of the sessions for the symposium. We would like to acknowledge all the supporters and organizers of this conference who shared their time and expertise. Finally, we would like to take this opportunity to thank the publication committee for their extensive support and strenuous efforts toward publication.

Contents

About the Editors

Dr. Adamu Murtala Zungeru is a Senior Lecturer in Electronics and Telecommunication Engineering and a Researcher at the College of Engineering and Technology at the Botswana International University of Science and Technology (BIUST). He obtained a Ph.D. degree in Electronics and Communication Engineering from the University of Nottingham, UK, and was a Research Fellow at the Massachusetts Institute of Technology (MIT), USA, where he also obtained a postgraduate teaching certificate in 2014. Before joining BIUST in 2015, he was a Senior Lecturer and Head of the Department of Electrical and Electronics Engineering at Federal University Oye-Ekiti, Nigeria. He is a Registered Engineer with the Council for the Regulation of Engineering in Nigeria (COREN) and the Association for Computing Machinery (ACM), USA, and a Professional Member of the Institute of Electrical and Electronics Engineers (IEEE), IEEE Antennas and Propagation Society (IEEE AP-S), and the Nigerian Society of Engineers (NSE). He is the Inventor of Termite-hill routing algorithm for wireless sensor networks and has filed two international patent applications with the Patent Cooperation Treaty (PCT) for his innovative work. He has authored 3 books and more than 40 international research articles and has over 10 years of university teaching and administrative experience. He has also served as an international reviewer for IEEE Transaction on Industrial Informatics, IEEE Sensors, JNCA, Wireless Networks, IET Networks, Sensors MDPI, Simulation-SAGE, IJCS, the Journal of Sensors-Hindawi, and numerous others.

S. Subashini is an Associate Professor at the School of Electronics at VIT University, where she has been working since 2012. She received a bachelor's degree in Engineering from Madurai Kamaraj University. She received her master's degree in Engineering from Anna University and her Ph.D. in Information and Communication Engineering from the same university in 2013. From 2004 to 2009, she worked at St. Joseph's College of Engineering, Chennai. Her research interests center on high-performance networks, cloud computing, security, and privacy. She has made significant contributions in the field of networking protocol design and analysis.

Dr. P. Vetrivelan is Associate Professor and former Programme Chair for Bachelor of Technology (Electronics and Communication Engineering) in School of Electronics Engineering at Vellore Institute of Technology (VIT), Chennai, India. He has completed Bachelor of Engineering from the University of Madras, Chennai, and both Master of Engineering in Embedded Systems Technologies and Doctor of Philosophy in Information and Communication Engineering from Anna University, Chennai. He has 14 years of teaching experience altogether in CSE and ECE Departments in both private engineering colleges in Chennai (affiliated to Anna University, Chennai) and a private engineering university in Chennai, respectively. He has authored 3 book chapters and one proceeding in lecture notes published by reputed Springer Nature publisher and has authored 25 Scopus indexed Journal papers and few other papers published in reputed international conferences. He has served as member in Board of Studies, doctoral committee, doctoral thesis examiner, and doctoral oral examiner in both private and government universities. He also serves as reviewer for reputed international journals and international conferences. His research interests include wireless networks, adhoc and sensor networks, VANETs, embedded systems, and Internet of Things (IoT).

Optimal Energy Saving Through Joint Deployment of Relay Station and Sleep Mode Activation in 4G LTE-A Network

R. Ratheesh and P. Vetrivelan

Abstract The advancement of information and communication technology (ICT) facilitates high-speed data rate centric real-time applications including streaming of video and instant chat. More number of macrobase stations (BSs) are required to accommodate fast growing number of users which constitute for a radical increase in energy consumption of the network. In order to mitigate this problem, a sleep mode algorithm for evolved NodeB (eNBs) of LTE-A networks with simultaneously powering relay stations (RSs) is the key trends in 4G networks. The sleep mode algorithm for base station is mathematically modeled with set theory. In proposed system model, each eNB is interconnected through X2 links and with RSs deployed in the transmission areas of selected eNB. The real-time network traffic of base station is estimated by a two-way handshake process between eNBs and mobile station. The control server (CS) is placed in the selected eNB which will decide to activate the sleep mode for RSs and eNB based on the estimated real-time network traffic profile. The comparison of optimal power consumption of BSs with RS is extensively simulated. The performance of sleep mode algorithm considering temporal variations of real-time network traffic is validated on hourly based scenario using MATLAB R2013a. The simulation results of the proposed work prove that there is an enormous power saving per eNB per hour. This proposed scheme is well suited for suburban area with temporal variations of network traffic.

Keywords eNB · Relay stations · Traffic profile · X2 links · Control server Sleep mode

---

R. Ratheesh (✉) · P. Vetrivelan
School of Electronics Engineering, VIT University,
Chennai Campus, Chennai 600127, India
e-mail: ratheesh.r2014@vit.ac.in

P. Vetrivelan
e-mail: vetrivelan.p@vit.ac.in

A. M. Zungeru et al. (eds.), *Wireless Communication Networks and Internet of Things*, Lecture Notes in Electrical Engineering 493,
https://doi.org/10.1007/978-981-10-8663-2_1

1 Introduction

With the evolution of information and communication technology (ICT) and the rapid increase of massive data traffic, energy consumption of the network is exponentially increasing. The implementation of 3G systems and emerging 4G technology in developing counties have significantly contributed to the development of information and communication systems of the nation but undesirably increased the power consumption of the network. Several studies indicate that within telecommunication networks, the wireless access networks are high-power consumers. Therefore, a lot of effort has been put lately in designing new power-reducing techniques such as sleep modes, cell zooming. Sleep modes allow a BS (or a part of BS) to switch off or put to power-saving mode when there is only a very less activity in its coverage cell. Whenever necessary, the BS is wakened up. When applying cell zooming, the cell size is adjusted adaptively according to the level of activity in a BS area. These techniques significantly reduce the power consumption in wireless access networks. In wireless systems, the high-energy consumption of a wireless base station (BS) results in noneconomical, large value of electricity bill. Greater than 50% of the total energy in wireless network is consumed by the radio access part, whereas 50–80% is spent for the power amplifier (PA) [1]. In [2], it is also mentioned that the energy bill accounts the Operation Expenditure (OpEx) for around 32% in India and roughly 18% in the mature European market. Another important motivation for power optimization in wireless networks is environmental awareness. Many of the base stations (BS) in rural areas which are not connected to power grid are powered by diesel generators for complete day and night as well as backup power source for few hours per day in urban and suburban areas. These diesel generators consume huge amount of diesel and emit large quantity of CO_2, which is a greenhouse gas (GHG). Three percentage [3] of the total CO_2 emission is from information and communication technology (ICT) industry throughout the world by consuming 2–6% [4] of the total worldwide energy. In operators view, the energy efficiency (EE) of wireless network not only brings ecological advantage and social benefits by solving issues for climate change but also has substantial economic benefits too [5]. With the exponential growth of large data transfer, it is unblemished that the ICT sector will become a major CO_2 emission sources within the next few years. Large energy consumption combined with its adversative effects on climate and environmental changes results in the need for an innovative energy-saving method for future. The remaining section of the paper is organized as follows. In Sect. 2, we discussed existing techniques for power optimization of radio BS from relevant literature. In Sect. 3, we described our proposed system of joint deployment of RS and BS. In Sect. 4, we present our proposed online network traffic calculation method. In Sect. 5, our relay-assisted BS power-saving algorithm (RABPS algorithm) is discussed. In Sect. 6, we showed the relevant result to evaluate our proposed work and concluded with Sect. 6.

2 Literature Survey

Many international research projects like Green Radio [6], EARTH [7, 8], OPERANet [9, 10], and eWIN [11] which outline the main solutions to energy efficiency in wireless communications are being carried out internationally. The network traffic is uncertain, but it is observed by practical studies and evident from various literature that there is a periodic peak hour and off-peak hour of network traffic daily and off-peak hours during holidays and weekends [12]. Figure 1 [12] shows uncertain traffic pattern with peak and off-peak hours. Cell shaping techniques such as cell breathing with respect to network traffic and hence activating sleep modes are the efficient method to optimize power of radio base station [13]. Cell zooming concept and sleep mode activation by calculating network traffic in terms of total power load is discussed in [14] and an algorithm for sleep mode activation is also discussed. Throughput optimization and capacity enhancement are described in relay-based heterogeneous network [15].

3 Proposed System of Joint Deployment of Relay Station and Base Station

The LTE-Advanced standard has specified the usage of relay nodes (RNs) as a cost-efficient means to extend the capacity of a base station (termed eNB, evolved NodeB). Each RS accesses the eNB through a wireless backhaul link (BL).

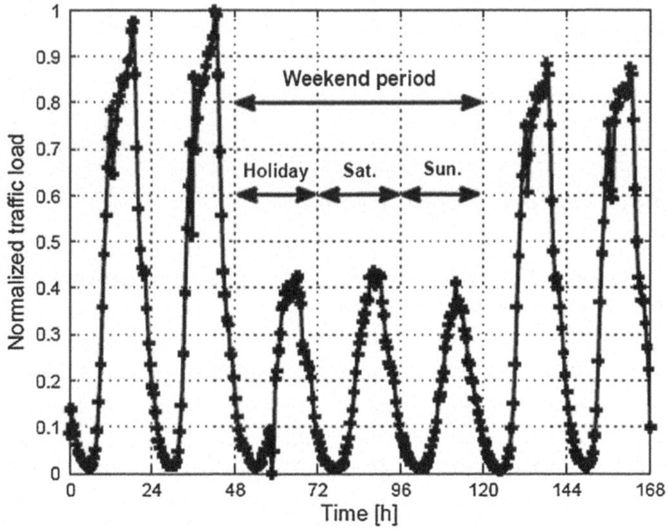

Fig. 1 Uncertain traffic pattern of BS during peak and off-peak hours

It forwards data to and from some user equipment (UE) through a wireless access link (AL). In this paper, we have proposed a centralized algorithm to estimate the number of MSs associated to each RS and the selected BS, by a two-way handshake process between MS and RS. The control server placed at BS computes power-saving algorithm and activates sleep mode to BS.

3.1 System Model

The BSs in the proposed work are evolved NodeB (eNB) which is capable of communicating with neighboring BSs and RSs which are deployed in the transmission range of selected BS through X2 links. The number of RSs to be deployed depends on the total coverage area of selected BS considering minimum overlapping area between them. The control server (CS) placed at the BS collects the information of number of MSs associated to each relay station through X2 link and computes the algorithm to make RSs and BS to sleep mode. Figure 2 shows the proposed system model with joint deployment of RS and BS. The control server also takes part in handoff management process between the RSs deployed in the selected BS.

4 Network Traffic Estimation

The implementation of sleep modes or power-saving modes for BS requires an efficient estimation network traffic. Various existing methods to estimate network traffic profile, such as prediction algorithms, probabilistic analysis are off-line traffic estimation methods. These methods use old data to predict the present network traffic.

4.1 Online Network Traffic Estimation

In order to activate sleep modes for BS, an efficient real-time network traffic profile calculation is required. In this work, we propose a method to estimate real-time network traffic based on the number of mobile stations associated to each RS deployed in the cell area of BS by a two-way handshake process between MS and RS. The control server (CS) placed at the BS collects information about all the users (MSs) which are associated to the corresponding RS deployed in the transmission area "*A*" of selected BS through the X2 link. The process flowchart for relay station is described in flowchart shown in Fig. 3. The control server (CS) computes an iXi

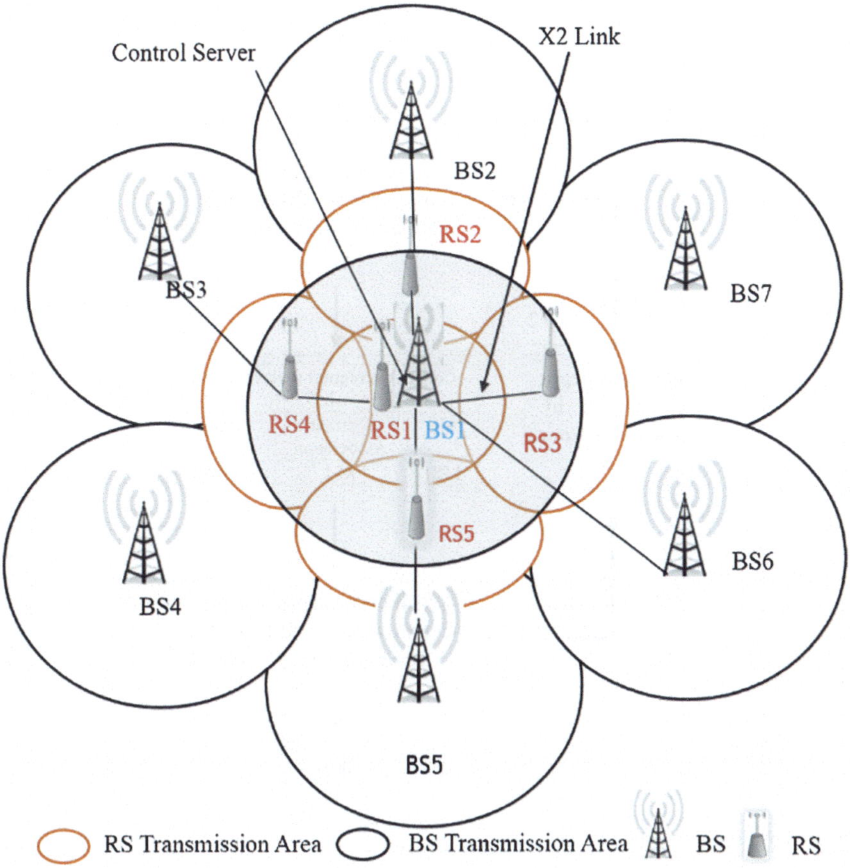

Fig. 2 System model with joint deployment of BS and RS

matrix (MS association matrix) with the collected data which are shown in Table 1, where i = number of RSs deployed within the transmission area of selected base station BSi. A mobile station (MS) will be associated to only one relay at a time, but from the data we received (refer Fig. 3) from each MS, we can find the next nearby RS also with respect to the signal strength. This information is used to fill remaining fields of the iXi matrix (MS association matrix) which will be helpful to provide service to particular MS while predicting mobility and handoff. Total number of mobile stations "M" in the area "A" of BSi is equal to sum of diagonal elements.

$$M = \sum \mathrm{RS}_{ij}, \quad \text{where } i = j. \tag{1}$$

If the total number of active MSs calculated is below the threshold value "T" of the total capacity of selected base station BSi, then the centralized algorithm to

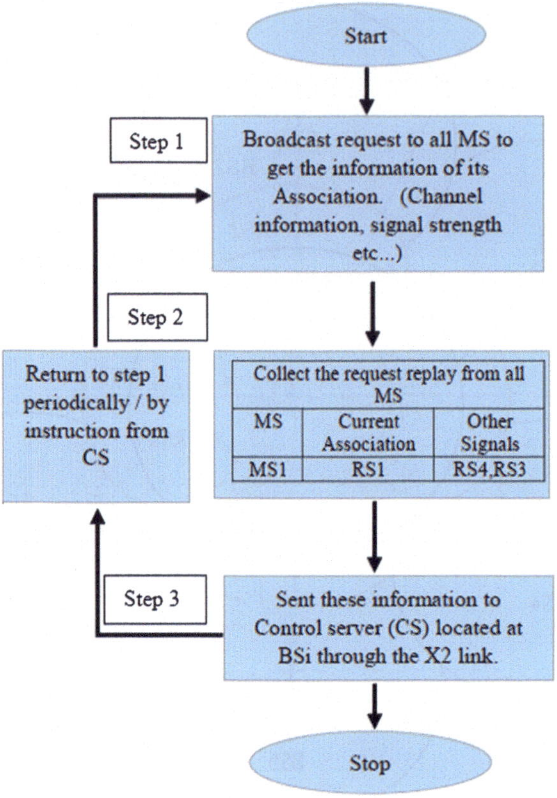

Fig. 3 Process flow graph at relay station

Table 1 iXi MS association matrix

	RS1	RS2	RS3	RS4
RS1	12	1	2	3
RS2	1	0	5	6
RS3	2	5	22	8
RS4	3	6	8	28
RS5	4	7	9	10

$$= \begin{pmatrix} 12 & 01 & 02 & 03 & 04 \\ 01 & 00 & 05 & 06 & 07 \\ 02 & 05 & 22 & 08 & 09 \\ 03 & 06 & 08 & 28 & 10 \\ 04 & 07 & 09 & 10 & 18 \end{pmatrix}$$

make BS to sleep mode is computed. Now if any one diagonal element is "0," it means that there is no MSs associated to that particular RS, so for further power consumption optimization, this relay station RS can be put into power-saving/sleep mode by zooming coverage area of neighboring RSs.

4.2 *Off-Line Network Traffic Estimation*

The off-line network traffic estimation methods use historically calculated data to predict the present network traffic.

4.2.1 Erlang Formula

An Erlang (E) is a unit of telecommunications traffic measurement as a measure of offered traffic. Offered traffic (in Erlangs) is related to the call arrival rate, λ, and the average call-holding time (the average time of a phone call), h.

$$E = \lambda h \tag{2}$$

The practical measurement of traffic is typically based on continuous observations over several days or weeks, during which the instantaneous traffic is recorded at regular, short intervals (such as every few seconds). These measurements are then used to calculate a single result, most commonly the **busy hour traffic** (in Erlangs). These practical results for several days are used to predict the network traffic in the near future, and with this predicted result and equating with the threshold T1, power-saving mode is activated.

Erlang B formula: This is the most commonly used traffic model and is used to work out on how many lines are required if the traffic figure (in Erlangs) during the busiest hour is known. The model assumes that all blocked calls are immediately cleared.
Erlang C formula: This model assumes that all blocked calls stay in the system until they can be handled. This model can be applied to the design of call center staffing arrangements where, if calls cannot be immediately answered, they enter a queue.

5 Relay-Assisted BS Power-Saving Algorithm (RABPS Algorithm)

The objective of this work is to save power of the network by putting the BS into power-saving mode (PSM) and simultaneously powering low-power RSs to ensure quality of service. From the iXi matrix (MS association matrix) computed, the value of each diagonal element gives the number of MSs associated with each relay. If any diagonal element is zero means, there is no MS associated with the particular RS. To save further power consumption, a sleep mode is activated in these RSs until the value changes by next iterative phase. The coverage area of RS which is in sleep mode is covered by neighboring active RSs by zooming their transmission area (cell boundaries) called relay zooming. This is done similar to a cell breathing technique. After ensuring total overlapped coverage for selected BS transmission area by powering remaining RSs deployed, sleep mode is activated to BS until the next iteration phase.

5.1 RABPS Algorithm

- Check for value zero in the diagonal elements of iXi (MS association matrix) matrix
- Identify the RSs which are not associated to any MS and put those RSs to power-saving mode/sleep mode
- Let "*X*" be the total number of RSs deployed, and let "*Y*" be the number of RSs which are not associated with any MS. Compute Z = number of active RSs

$$Z = X - Y \tag{3}$$

- Let R_X, R_Y, and R_Z be the set of total number of RSs deployed, RSs in sleep mode, and RSs in active mode, respectively.
- Start with an element R_{Yi} of the set R_Y which is a relay station not associated to any MS.
- Determine its closest neighbors, which are the elements of set R_z and form a set R_m where $R_m = \{R_j\}$, where $j = 1$ to M, M = the number of closest RSs of $R_{Yi} \in R_Y$, where $i = 1$ to "*Y*"
- Zoom the transmission area of determined neighbors from $R_m = \{R_1, R_2 \ldots R_M\}$ one by one.
- Check weather transmission area of R_{Y1} is overlapped.
- If no, $j \leq M$, j++
- If yes, continue with next element of R_Y by incrementing $i = i + 1$, $i \leq Y$, return to step 5.
- Check $i = Y$,
- If yes, check for solution, solution is total coverage of selected BS area by activating RS and activate sleep mode for selected BS.
- Else, continue in normal mode.

6 Results and Discussion

A network topology with nine relay stations (RSs) deployed in a BS transmission area with minimum overlapped coverage and with random distribution of mobile stations (MSs) were simulated using MATLAB 2013a 8.1 version, which is shown

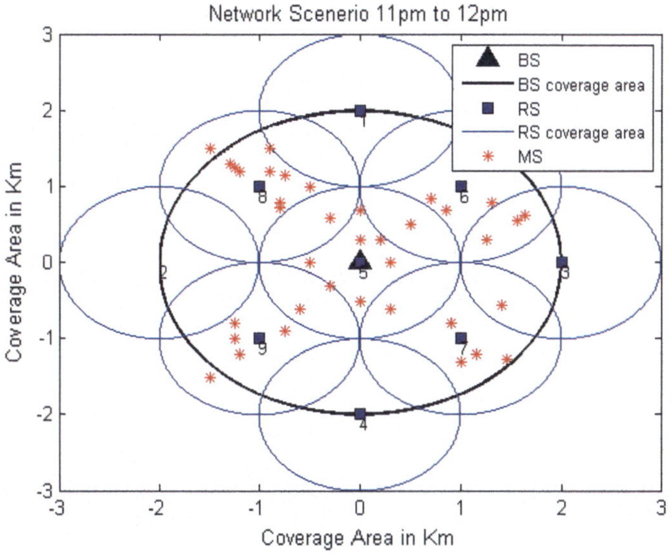

Fig. 4 Network scenario for a off-peak hour from 11 pm to 12 pm

in Fig. 4. A suburban region with temporal variation of network traffic is considered. Figure 5 shows the network traffic profile for a day. A threshold below 50% of total capacity is assumed as off-peak hours. Power consumption comparison for one hour between 11 pm and 12 pm is done in simulation.

The total number of mobile stations in the transmission area of BS is calculated by summing the diagonal elements as mentioned in Sect. 3, by Eq. (1). The RSs with none of the MS associated with it is calculated by identifying the RS corresponding to "0" diagonal elements of iXi matrix (MS association matrix). The power rating of the selected BS is assumed as 80 W/h, and the power rating of relay station is 8 W/h. The transmission area of BS and each RS is assumed as 2 and 1 km, respectively. So for covering the transmission area of selected BS nine RSs are needed to deploy. The traffic profile of the selected BS is considered to be in off-peak hours from 11 pm to 7 am.

With the random distribution of MSs in the network scenario shown in Fig. 4, the RS 1, RS 2, and RS 4 are not associated with any MS. Power-saving mode is activated to these relay stations until the next iteration phase for time "*t*." In order to activate sleep mode for RS 1, RS 2, and RS 4, its area of coverage is to be covered by neighboring active RS 6, RS 7, RS 8, and RS 9. After performing RABPS algorithm, the sleep mode for BS is activated for time "*t*." The power consumption analysis of base station with and without the proposed RABPSM algorithm is done for one hour. During off-peak hours for normal operation of BS, it is assumed to consume 80 W per hour. Simulation of RABPS algorithm shows that for one hour of off-peak neglecting transition time power, 30 W power is saved for this typical scenario, which is shown in Fig. 6. For the real-time traffic analysis, both the RSs

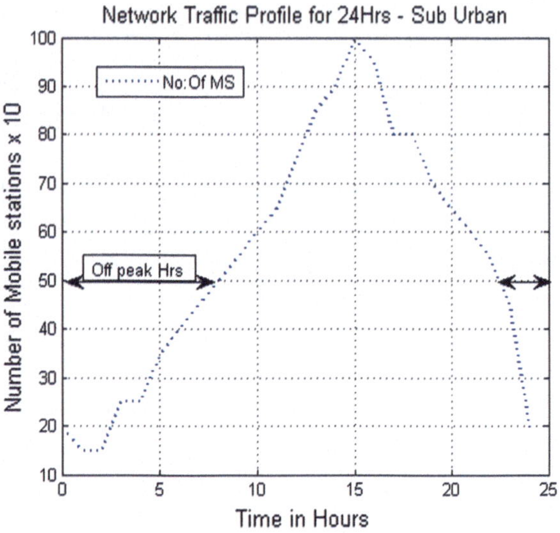

Fig. 5 Network traffic profile for a day

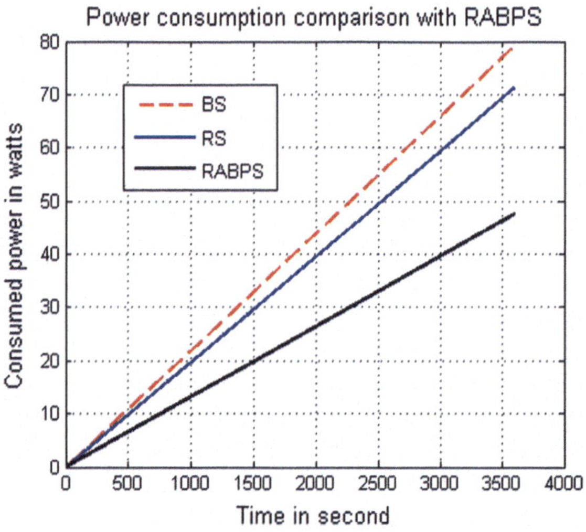

Fig. 6 Total power consumption of BS with and without RABPS algorithm

deployed and BS are needed to power together for few seconds. But after the network traffic analysis, the RSs which are not associated to any MSs are made to sleep. The power consumption of the total system during this initial phase is higher for few seconds which is transition period power consumption shown in Fig. 7. In this simulation, the simulation running time is considered as transition time of the system.

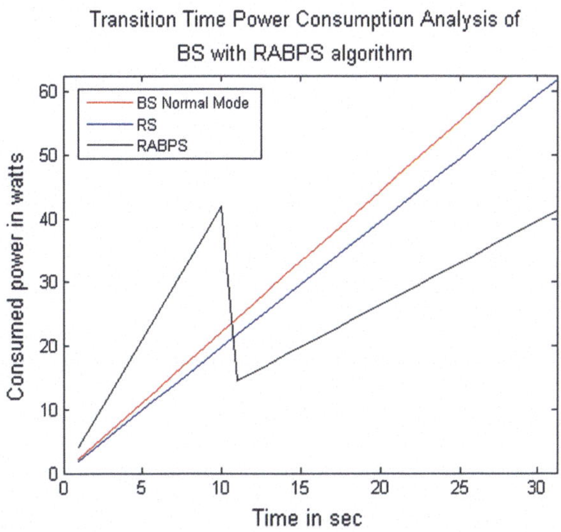

Fig. 7 Transition time power analysis of RABPS algorithm

7 Conclusion

In this paper, we proposed a novel RABPS algorithm to optimize power consumption of radio base station by activating sleep mode and simultaneously powering RSs deployed in the cell. We also proposed a method for real-time network traffic estimation which can efficiently find out the number of MSs associated to each RS and the BS. The concept of relay zooming is introduced in this work to cover area of transmission of adjacent RSs, which are in sleep mode. The simulation results show that our proposed method can save an enormous amount of power for radio base stations without affecting quality of service. The deployment of RSs on the cell edges is an efficient methodology to manage group handoff, which will be focused in our future work. The sleep mode algorithms for BS not only reduce power consumption of BS but also help to reduce the emission of GHG gases from BS. The frequency planning for RSs deployed inside a cell needs to be done, as spectrum is limited, which will be focused in our future work.

Acknowledgements We would like to thank the anonymous reviewers for their comments for improving this paper, and also, we extend our gratitude to VIT University, Chennai, for their support.

References

1. Suarez L, Nuaymi L, Bonnin J-M (2012) An overview and classification of research approaches in green wireless networks. EURASIP J Wirel Commun Netw 2012:142
2. Lister D (2009) Vodafone group research & development, "An operator's view on green radio". Presented at the proceedings of IEEE international workshop on green communications

3. Militano L, Molinaro A, Iera A, Petkovics A (2012). Introducing fairness in cooperation among green mobile network operators. In: Proceedings of 20th international conference on software, telecommunications and computer networks (SoftCOM), Italy, pp 1–5
4. Lambert S, Van Heddeghem W, Vereecken W, Lannoo B, Colle D, Pickavet M et al (2012) Worldwide electricity consumption of communication networks. Opt Express 20(26): B513–B524
5. Ratheesh R, Vetrivelan (2016) Power optimization techniques for next generation wireless networks. Int J Eng Technol (IJET). e-ISSN: 0975-4024
6. Grant P (2010) MCVE Core 5 Programme, "Green radio-the case for more efficient cellular basestations". Presented at the Globecom'10
7. EARTH. Energy aware radio and network technologies project. https://www.ict-earth.eu/default.html
8. Most promising tracks of green network technologies (2010) INFSO-ICT-247733 EARTH Deliverable D3.1, Earth, WP3-Green Networks. https://bscw.ict-earth.eu/pub/bscw.cgi/d31509/EARTHWP3D3.1.pdf
9. OPERA-Net, Optimising power efficiency in mobile radio networks project. http://opera-net.org/default.aspx
10. Optimising power efficiency in mobile radio networks (2010) OPERA-Net PROJECT STAND #42, 2010 NEM Summit Towards Future Media Internet, Barcelona, Spain, Oct 2010. http://opera-net.org/Documents/5026v1Opera-Nete-NEM%20Event%20Barcelona%202010Demos%20Presentation290710.pdf
11. Energy efficiency enhancements in radio access networks (2008) Wireless@KTH Research Strategy Document 2008–2010, Wireless@KTH. http://www.wireless.kth.se/images/stories/Strategy/Researchplan08.pdf
12. Saxena N, Sahu BJR, Han YS (2014) Traffic-aware energy optimization in green LTE cellular systems. IEEE Commun Lett 18(1):38–41
13. Micallef G, Mogensen P, Scheck H-O (2010) Cell size breathing and possibilities to introduce cell sleep mode. In: 16th European wireless conference 2010, Lucca, Italy, pp 111–115
14. Deruyck et al (2012) Characterization and optimization of the power consumption in wireless access networks by taking daily traffic variations into account. EURASIP J Wirel Commun Networking
15. Ouni A, Saadani A, Rivano H (2013) Energy and throughput optimization for relay based heterogeneous networks. IEEE. ISBN: 978-1-4799-0543-0/13/$31.00

Efficient Relaying for Enhanced Network Longevity for E-health IOT Services in Medical Body Area Networks

R. Latha and P. Vetrivelan

Abstract *Background* Network lifetime is an essential performance metric for medical body area networks (MBAN) since nodes meant to monitor medical parameters continuously for a longer amount of time. This paper focuses on various schemes for enhancing the network lifetime through an analytical approach. *Methods* An analytical model for enhanced lifetime of sensor nodes is proposed. The proposed model consists of both relay nodes (RN) and sensor nodes (SN) towards enhancing the longevity of network lifetime. The routing protocol is designed to collect the sensed information from SN by adapting optimal path towards gateway. The optimal distance between SN and RN is chosen based on best quality link which ensures the packet delivery. The total installation cost of RN, the total energy consumption of both RN and SN, the traffic serviced from all sensors and data routing to nodes are the constraints considered in the proposed framework for achieving the optimal path though efficient relaying. *Results* The proposed multi-tier telemedicine system is extensively simulated for describing the optimal network path for MBAN-based e-health services. The proposed work is modelled using integer linear programming that optimizes the location and number of relays. Relays are deployed for minimizing the cost of network installation of RN. The energy which is consumed by both sensors and relays is minimized while ensuring full coverage with effective routing of e-health services. The network longevity is analyzed during both normal and emergency scenarios.

Keywords MBAN · Network lifetime · Installation cost · Energy
Relay nodes

R. Latha (✉) · P. Vetrivelan (✉)
School of Electronics Engineering, Vellore Institute of Technology,
Chennai 600127, India
e-mail: latha.r2015@vit.ac.in

P. Vetrivelan
e-mail: vetrivelan.p@vit.ac.in

A. M. Zungeru et al. (eds.), *Wireless Communication Networks and Internet of Things*, Lecture Notes in Electrical Engineering 493,
https://doi.org/10.1007/978-981-10-8663-2_2

1 Introduction

Medical body area network (MBAN) offers a reliable and comfortable platform for measuring the parameters like heart rate, blood pressure, blood glucose and walking rate of the respective patient and records them in their databases. With the help of this technology, continuous monitoring is possible. Some of the most important requirements of MBAN are coexistence, low power, robustness and scalability. All the wearable sensors must communicate the information to a hub through which the information is transmitted to the doctor who monitors the patient data. These hubs are in turn connected to the central control point, which acts as an interface between hub and sensors for doing MBAN communication in 2360–2390 MHz as specified by FCC. This control point in addition receives data message for enabling the use of certain frequencies offered only for MBAN devices.

Power scarcity is an important issue in MBAN. For optimizing the sensor energy, there is an impact on extending the network lifetime.

In this paper, the network lifetime is enhanced through an integer linear programming model. The number of relay nodes (RN) and sensor nodes (SN) to be deployed for minimizing the network installation cost is found out.

These are the contributions of this paper: (1) mathematical model for enhanced lifetime of network, (2) design heuristics (both normal and emergency) using integer linear programming and (3) analyse the performance metrics through simulations.

The paper is as follows. Section 2 describes the related works on network lifetime enhancement. Section 3 gives the heuristics on integer linear programming techniques. Section 4 deals with the simulation results and Sect. 5 gives conclusions.

2 Related Work

In past, some works related to increase in network lifetime of WBAN using relay nodes [1, 2]. In [2], relaying was combined with cooperation for handling traffic and was not utilized for sensing; hence, more energy is available. In addition, the relay nodes' position is fixed but not optimized. In [1], the upper bound is specified for the number of relays. For prolonging the network lifetime, in [3] an optimized MAC protocol which is based on IEEE 802.15.4 standard for the communicating with gateway and for switching off from one hop to multi-hop topology. Multi-path routing protocol is used for WBAN in [4], which helps in increase of the network lifetime with the help of ADHOC mode for importing the mobility of the patient in the hospital. Using thermal energy harvesting from the human body, [5] explains in improving the network lifetime of nodes.

3 Integer Linear Programming Model

The mathematical optimization assumes that the variables as integers. There are integer linear programming (ILP), in which the objectives and constraints are linear and 0–1 integer linear programming, where the unknowns are binary. Problems modelled using ILP are definitely NP hard like that of travelling salesman problem. The integer variables express decisions and hence, they are much useful in analysing networks. In networks, the goals, like minimizing the installation cost, distributing the available frequencies to antenna, are taken into account.

3.1 Network Model

A WBAN scenario is considered where sensors are connected to gateway through relay nodes. The positions of the relays are not fixed, whereas the position of sensor nodes is fixed. Let SN be the sensor set, RN denote the set of relays and C denote coordinator/sink node which acts as coordinator for collecting all the sensed information and send to the server. Figure 1 shows the network model of MBAN. This diagram depicts how the SN and RN communicate the health data of a patient to the doctor through gateway and server.

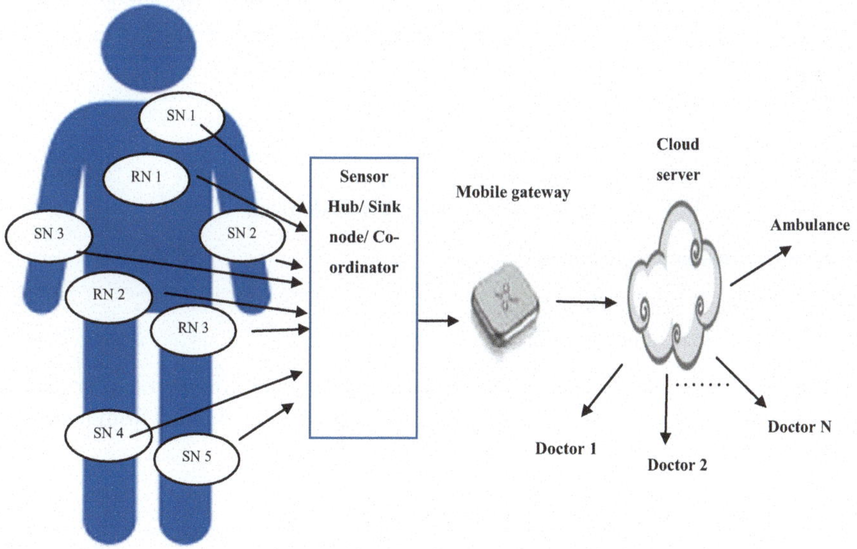

Fig. 1 Medical body area network with SN and RN sending information to the doctor through the mobile gateway

The objective is to minimize the distance between SN and RN. The network route is defined from SN to RN. The constraints taken are network installation cost of relay nodes, and the energy consumed by SN and RN must be minimized.

3.2 Propagation Model

The propagation model between the SN and RN is adopted from [5, 6]. The energy is calculated according to the energy consumed by SN, RN and sink node. In addition, sensing energy and processing energy are assumed to be less when comparing to the energy of communication. The energy consumption is the addition of transmission as well as reception energy of the wireless nodes.

3.3 Enhanced Network Longevity of MBAN Model

According to the above-mentioned parameters, an efficient relaying for enhanced network lifetime for e-health is designed as follows. The design specifications include that the distance between the SN and RN is minimized as well as the cost of network installation and the energy consumption of wireless nodes is minimized.

The objective function takes into account the distance between the SN and RN, total installation cost of the RN and the energy consumption of the wireless nodes. This optimization problem is solved by modelling as flow of units of a product through a network of nodes and connections between nodes.

For every node,

- Sink node/coordinator, $b_i > 0$, has demand of b_i units
- Sensor node/SN, $b_i < 0$, has supply of b_i units
- Relay node/RN, $b_i = 0$, has neither supply nor demand

Connections between nodes

- c_{ij} = cost per unit of flow from i to j
- l_{ij}, u_{ij} = lower, upper bounds

Figure 2 shows the network flow model through which routing of information takes place between source and destination nodes. x_{ij} = units shipped from node i to node j

$$\text{MIN} \sum_{ij} c_{ij} x_{ij} \tag{1}$$

$$\text{S.T.} \sum_{k} x_{ki} - \sum_{i} x_{ii} = b_i \tag{2}$$

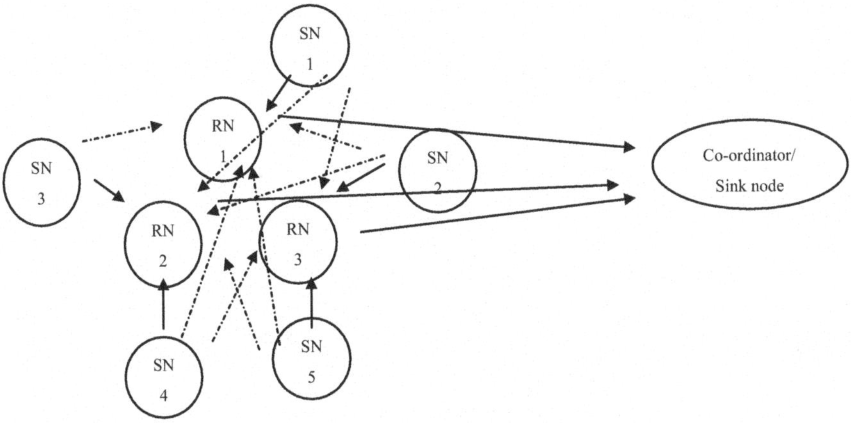

Fig. 2 Possible routes for data from SN to RN for reaching the sink node

$$\sum_k x_{ki} - \sum_i x_{ii} \le e_i \tag{3}$$

where e_i is the energy consumption of wireless nodes.

4 Numerical Results

This section explains the proposed enhanced network longevity with the different parameters like the number of SN, RN as well as the values in objective function, which contributes to the cost-effective and energy efficient network.

Whenever there is a sensed information, the SN routes them from SN to the sink node/coordinator through the RN. Here we have assumed that 5 SN and 3 RN are available. Table 1 shows the assumed route relations from SN to RN with the constraints assumed such that the network installation cost is minimum (i.e. 100) and the energy consumption of the wireless nodes to be lesser than or equal to 300.

Table 2 shows the control limits of the rate of data packets, which helps in determining the minimization of the objective function.

Figure 3 shows the change in objective value w.r.t. the data rate. The minimum objective value is obtained when all the data rates are equal to 1 kbps.

Table 1 Routes from SN to RN and the network cost and energy constraints

Nodes	To			Objective value	500	Network cost
From	RN 1	RN 2	RN 3			
SN 1	25	0	75	100	=	100
SN 2	25	0	75	100	=	100
SN 3	0	100	0	100	=	100
SN 4	25	0	75	100	=	100
SN 5	25	0	75	100	=	100
	100	100	300			
	≤	≤	≤			
Energy	300	300	300			

Table 2 Control of data rates from SN to RN

Nodes	To		
From	RN 1	RN 2	RN 3
SN 1	1	1	1
SN 2	2	2	2
SN 3	3	3	3
SN 4	4	4	4
SN 5	5	5	5

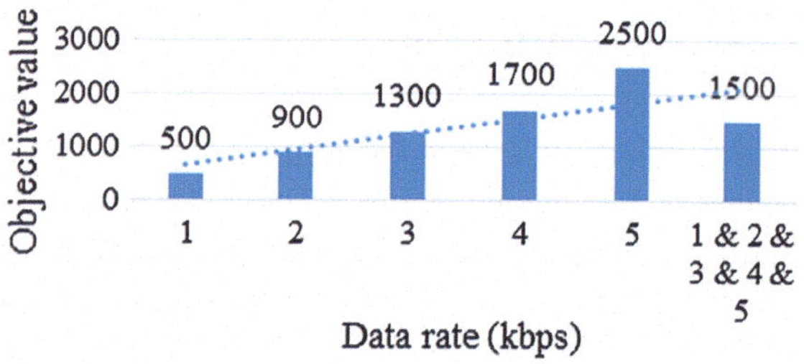

Fig. 3 Variation of objective value w.r.t. data rate

5 Conclusion

Thus a multi-tier telemedicine system is stimulated for describing the optimal network path for MBAN-based e-health services. This was modelled using linear integer concept for optimizing the location and relays that are to be deployed for minimizing network installation cost of RN.

Acknowledgements The authors would like to thank the Institute Professors for helping to improve this study. The authors also thank the reviewers for giving input on this study.

References

1. Ehyaie A, Hashemi M, Khadivi P (2009) Using relay network to increase lifetime in wireless body area sensor networks. In: Proceedings of the 10th IEEE WoWMoM, Kos, Greece, June 2009, pp 1–6
2. Reusens E, Joseph W, Latre B, Braem B, Vermeeren G, Tanghe E, Martens L, Moerman I, Blondia C (2009) Characterization of on-body communication channel and energy efficient topology design for wireless body area networks. IEEE Trans Inf Technol Biomed 13(6): 933–945
3. Azzouz BB, Mohamed B, Abdesselam B, Goursaud C, Hutu F (2015) Enhancement optimized MAC protocol for medical applications. In: New technologies of information and communication (NTIC), 2015, pp 1–6
4. Birgani YG, Javan NT, Tourani M (2014) Mobility enhancement of patients body monitoring based on WBAN with multipath routing. In: Information and communication technology (ICoICT), 2014, pp 127–132
5. Ghosh A, Khalid S, Harigovindan VP (2015) Performance analysis of wireless body area network with thermal energy harvesting. In: Global conference on communication technologies (GCCT), 2015, pp 916–920
6. Heinzelman WR, Chandrakasan A, Balakrishnan H (2000) Energy-efficient communication protocol for wireless microsensor networks. In: Proceedings of the 33rd annual Hawaii international conference on system sciences, MIT, Cambridge, MA, USA, 2000, pp 3005–3014

Gradient-Based Localization and Relay Nodes Selection in Delay Tolerant Mobile Opportunistic Networks for Emergency Rescue

C. P. Koushik and P. Vetrivelan

Abstract *Background* In opportunistic networks, there is no complete path between source and destination and all the nodes are in mobility. Opportunistic networks exploit the potential capability of existing mobile nodes, such as data forwarding without any preplanted infrastructures. In gradient-based routing, relay nodes are selected using a gradient value that leads to less assurance to reaching the destination. So finding the location of the relay node is an important criterion, which resolves to loop the packets within the network. Localization is one of the critical issues for handling emergency rescue message in opportunistic networks. *Method* In gradient-based localization, choose the best neighbor node that is near to the sink as a relay node. Nodes are two types in the network; they are anchor nodes and sensor nodes. Anchor node is equipped with GPS receiver, and sensor nodes are randomly deployed in canvas. Anchor position is considered as a reference, and in turn sensor nodes position is computed using a hybrid approach. Hybrid approach maps the anchor node coordinates, ToT (time of transmission), AoA (angle of arrival), and ToA (time of arrival) into a Cartesian plane in order to localize the sensor node position. After finding the sensor position, the relay node selection was based on near to the sink, which maximizes the emergency message discrimination toward the sink. *Results* Proposed localization and relay node selection were extensively simulated in OMNeT++. The simulation profile consists of 20 anchor nodes and 30 sensor nodes in the considered terrain dimension. The simulation result shows that the performance metric such as a Packet Delivery Ratio, throughput, and End-to-End Delay is enhanced and compared with existing gradient-based approach. This proposed scheme is most suitable for dissemination of a rescue message with least delay during emergency cases such as floods, volcano, and earthquake scenario.

C. P. Koushik · P. Vetrivelan (✉)
School of Electronics Engineering, Vellore Institute of Technology,
Chennai 600127, India
e-mail: vetrivelan.p@vit.ac.in

C. P. Koushik
e-mail: koushik.cp2014@vit.ac.in

A. M. Zungeru et al. (eds.), *Wireless Communication Networks and Internet of Things*, Lecture Notes in Electrical Engineering 493,
https://doi.org/10.1007/978-981-10-8663-2_3

Keywords AoA · Gradient-based routing · Localization · ToA
ToT

1 Introduction

In Mobile Ad hoc Network (MANET), nodes move in random way mobility model which is irregular in motion. It is difficult to get the topology information with it, and it is an infrastructure-less network. A topology change in MANET also occurs due to frequent change in network and unstable network connection, energy and lack of infrastructure. These are the major issues in MANET compared to other network [1]. MANET finds varied applications. Some of them are: tactical network, emergency service, home network, and context-aware network, etc. [2]. In the writing, discontinuously associated networks are regularly indicated to as *delay*- or *disruption tolerant networks* (DTN) be that as it may this term is all the more entirely connected with the Delay/Disruption Tolerant Networking design [3]. Delay Tolerance Mobile Sensor Network [4] is a type of sensor network, which is technically origin from Delay Tolerance Network (DTN). DTMSNs have several unique characteristics like mobility of nodes, density of networks, and buffer space is limited and short radio communication. Since this characteristic leads to disconnection in network and frequent changes in network topology, it is rarely possible to have a complete path between source and destination and there will be frequent change in network topology in DTMSN.

Initially while deployment each sensor nodes location is given to them. This id is done either manually or using GPS devices attached to sensor nodes. Installing GPS in sensor nodes is not possible in the context of large network because of the excessive cost and workforce involving, respectively. To overcome this, sensor nodes are made to identify their location with the help of neighbor nodes [5]. This paper focuses on the localization technique to find the location of neighbor node or non-GPS-enabled nodes. Several researches are going on in this field of localization to route the packet to sink.

2 Related Work

Spano and Ricciato [6], a planner (2D) scenario with Mnunber of nodes and M-1 anchor nodes to find the position of that one blind node. In general scenario, every node transmits a packet to find the distance of unknown transmission time by observing period of duration *D*. Here, every packet is transmitted over air as part of the normal communication process to exploit for localization purpose: Data packet, ack, and periodical beacons are used to find the ToA (time stamp). All the nodes are capable of transmitting (*Tx*) and receiving (*Rx*), but here dose not required necessary fully-meshed communication i.e. it can be applied to partial-mesh communication.

Author follows the same approach adopted by GPS, where the problem is split into two stages. As packet time stamp and mutual distance between the anchor nodes are used to estimate the set of pseudo-ranges between the anchor nodes and blind nodes, then in the second stage, the blind node position is estimated from the set of pseudo-range combined with anchor node position.

Das and Thampi [7], Network architecture consists of two types of nodes: anchor node (*A*-node) which is GPS enabled and sensor node (*S*-node) is randomly deployed in the underwater domain. *A*-node periodically broadcasts a message, and *S*-node receives that message to track its location. The data available for an *S*-node to localize are *A*-node coordinates, time of transmission, AoA, and ToA. Once *S*-node successfully receives the broadcast packet from *A*-node, it immediately starts the location computation phase. Location computation phase depends on the successful reception of the packet from the *A*-node. Hence, equirectangular approximation is chosen to map the geographical coordinates to the *S*-node in the Cartesian plane. The coordinates of the *S*-node in the Cartesian plane are estimated and converted back to geographical coordinates. The range between the *A*-node and *S*-node is measured using ToA. Therefore, minimal storage space is required and communication overhead is less when compared to other existing positioning techniques.

Zhang et al. [8], if a mobile sensor node enters the unknown area, it must be able to detect its own location using GPS, alternatively a dynamic localization scheme that must also be used which adjusts the estimation location of the node based on recent motion. The main goal of gradient-based target equation scheme is to predict the localization of stationary target within an allowable uncertainty. The base station disseminates a search objective to mobile sensor network with two parameters, the error tolerance and the desired quality of the target. With this, an objective could be to localize the target within 2 m with at least 95% confident. Once the individual node has received a request from the target acquisition, each one determines the most efficient way to localize the potential target with the request level. In particular, anode starts moving in the direction, which it anticipates is the shortest path to reach the sink.

Boukerche et al. [9], author explains in detail about RSSI, ToA, and AoA in this paper. RSSI can be used to estimate the distance between two nodes based on the signal strength of the signal received by another node. Sensor node sends a signal with a determined strength that fades as the signal propagates. The bigger the distance to the receiver node, the lesser the signal strength when it arrives at the node. Radio propagation model can be used to convert the signal strength in distance. For ToA, distance between two nodes is directly proportional to the time the signal takes to propagate from one point to another. This way, if a signal was sent at time t_1 and reached the receiver node at time t_2, the distance between sender and receiver is $d = s_r(t_2 - t_1)$, where s_r is the propagation speed of the radio signal. This type of estimation required precisely synchronized nodes and the time at which the signal level the node must be in the packet that is sent. Using directional antenna or an array of the receiver does the estimation of the AoA; usually, three or more than three are uniformly separated. In the last case, based on the arrival times of the signal at each of the receiver, it becomes possible to estimate this signal.

The development of opportunistic applications, i.e., application running over opportunistic network, is still in early stage. This is due to lack of tools and supports the process in uncertain condition. Indeed, many tools have been introduced to study and characterize opportunistic network, but none of them is focused on helping developers to conceive opportunistic application. The gap between opportunistic application development and network characterization can be filled with network emulation [10]. Using the HINT network emulator to develop opportunistic chat application, which helps to test and performance evaluation perspectives. They show three different use cases for HINT: How to connect an application to an opportunistic network, to test the application using the HINT emulator, and finally to easily test different application scenarios using the HINT monitoring system [11].

2.1 Gradient-Based Routing

In gradient-based routing protocol, there is no complete path to transfer the message from source to destination due to random mobility of nodes. In DTMSNs, it will forward the data rather than flooding so it is difficult to gather topology information and to find the best path to route the message [12]. One promising way to deal with transferring the messages from source to sink on DTMSNs is to use gradient-based routing approach. In this segment, we quickly portrait the major thoughts of gradient-based routing.

2.2 Routing in Gradient-Based Approach

In gradient-based routing, every node has a metric through which we calculate how useful that node possible act as a relay node to transfer the message to sink node. Let us consider every node i has a gradient value G_i that has a packet to transmit to sink node; i.e., the destination node has a higher gradient value among all nodes in the network. When node i has neighbor node in its communication range, it check with its gradient metric is higher than its own it will transmit the message to its neighbor. Let us consider node j as one of the neighbors in node i whose value is G_j the best among its neighbor. Let us consider there is neighbor node with higher gradient value and whose value is higher than source node i then it will forward the message to node j or it will stores the packet in its own buffer. Only by using transmitted MSG along the gradient formed by G_i message can also be delivered at the sink node productively without the information of the whole network besides gradient metrics.

2.3 Need for Localization in Gradient-Based Approach

In gradient-based routing protocol, nodes have higher prospect to deliver the message by assigning higher gradient value for nodes near to sink. When the node having higher gradient value moves away from sink due to node mobility, then there is chance of packet transmitted within the network without reaching sink. This kind of approach cannot be trusted in emergency rescue services. Localization is the technique to find the localization of neighbor nodes and choose the neighbor node near to sink as relay node. By that we will choose the entire relay node that is near to the sink, so there is no chance of packet transmitted within network. Choosing best relay node near to the sink leads to reduce delay and reduce number of hop count for packet to reach sink.

3 Proposed Gradient-Based Localization Strategy

3.1 Cartesian Plane

The whole canvas is split into two numbered lines drawn perpendicular to one another, intersecting at zero on each number line. The horizontal number line is the *x*-axis, and vertical line is the *y*-axis. The two numbers divide the plane or canvas into four regions called quadrants. The point of intersection of the numbered line is origin as shown in Fig. 1. Each point in the plane is identified as an ordered pair (x, y) of real number called coordinates. Keeping sink node as origin (0, 0), Cartesian plane is drawn in canvas. Here, we are going to use two types of nodes: one GPS-enabled node or anchor nodes and another one is a normal sensor node. Anchor nodes frequently update its coordinated in the Cartesian plane with the help of GPS. Aim is to find the location of sensor nodes with the help of anchor nodes using Cartesian plane called network Cartesian plane.

3.2 Neighbor Localization

Each node in the network generates its own local Cartesian plane for its communication range as shown in Fig. 3, to find the location of its neighbors. Whenever a node has a packet to transmit, it will broadcast a RREQ message to its communication range to find the neighbor nodes and replay back with RREP message to source node. By keeping source node as origin neighbor nodes, Cartesian coordinates are calculated. RREQ/RREP has all the required information to calculate the neighbor coordinates as shown in Fig. 2.

Source node sends RREQ message to its entire neighbor node in its communication range. Constructing local Cartesian coordinates and network Cartesian

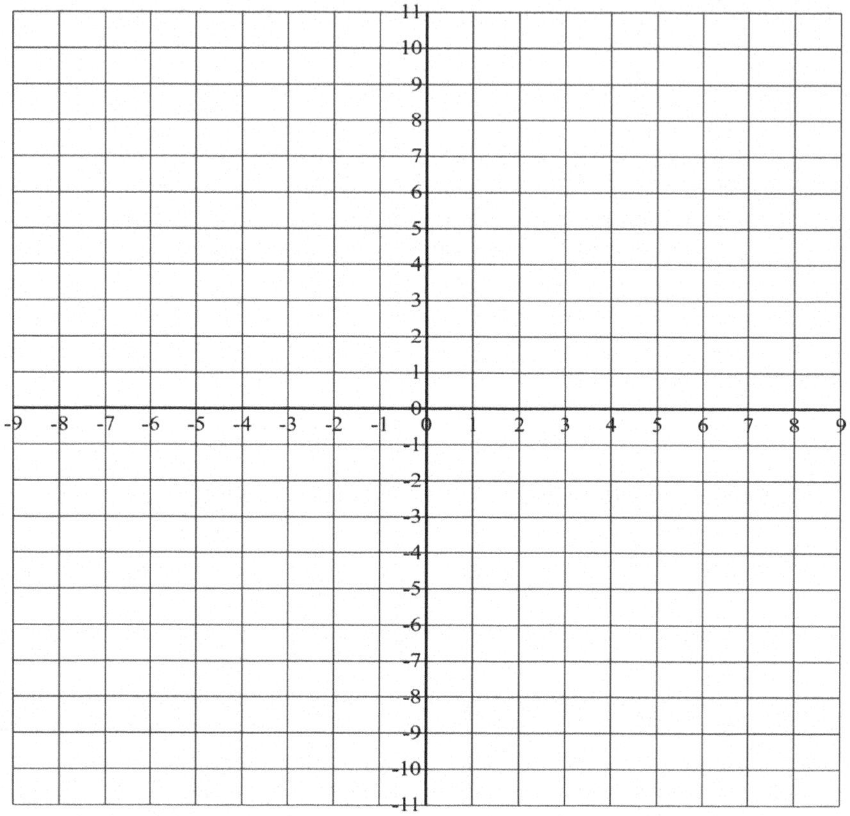

Fig. 1 Cartesian plane on canvas

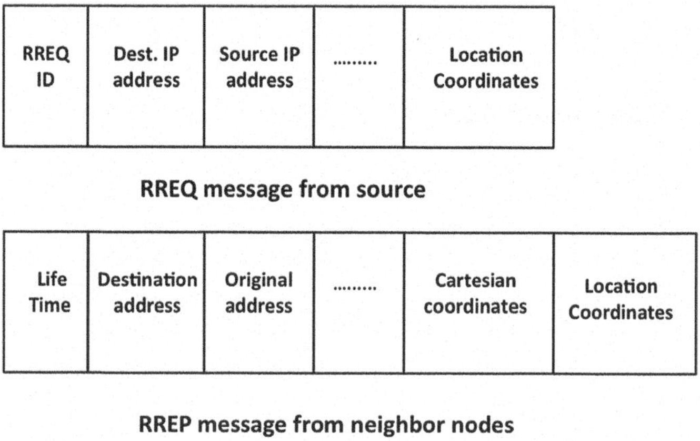

Fig. 2 RREQ/RREP message for communication

coordinates can identify nodes. Source will send its own location coordinates to its neighbor nodes if it is an anchor node or GPS-enabled node; if it is a sensor node, then in RREQ message location coordinates are empty. Nodes that receive RREQ from source will generate local Cartesian coordinate by keeping source as origin and replay back with its own coordinates in RREP message.

3.3 Calculating Cartesian Coordinate

To calculate the Cartesian coordinates of the nodes, first we have to find the distance "*d*" between the two nodes (*S*-node and *P*-node) using the RSSI value indicator. Using directional antenna, we can find angle (θ) of the node from which RREQ message is arrived. Then by using the vectors *i* and *j*, right-angled triangle is formed by moving the vectors *i*, *x*-unit parallel to *y*-axis and vector *j*, *y*-unit parallel to *x*-axis as shown in Fig. 3.

$$\sin\theta = x/d \rightarrow d\sin\theta = x \tag{1}$$

$$\cos\theta = y/d \rightarrow d\cos\theta = y \tag{2}$$

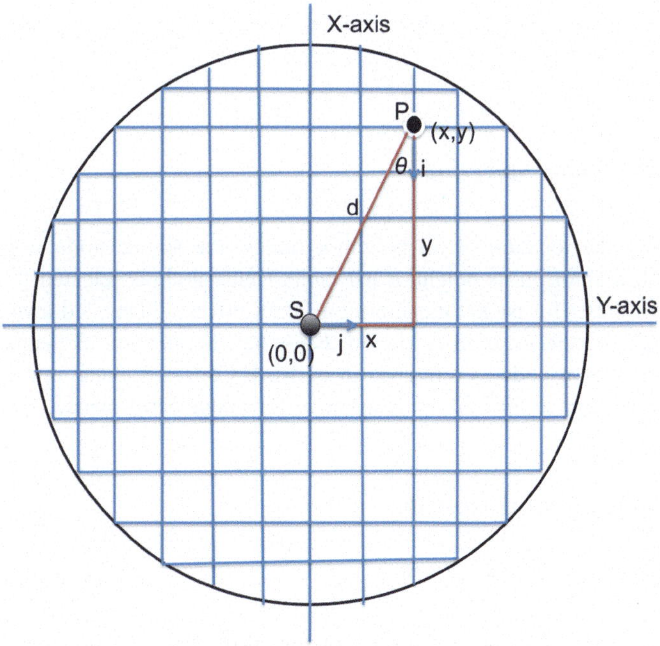

Fig. 3 Cartesian coordinate for single node

By substituting the distance "d" and angle of arrival (θ) in Eqs. (1) and (2) to get the value of x and y in right angular triangle, three possible conditions can occur in finding the location of neighbor nodes; they are:

1. Source node is GPS enabled: In this source node will send its location in RREQ message to its neighbor nodes, by keeping source node as reference node using Cartesian plane neighbor nodes will compute its location.
2. One of the neighbor nodes is GPS enabled: In RREP, anchor node will send its location along with Cartesian coordinate to source node. Source node will compute its location with RREP of anchor node.
3. No anchor node: Then source node cannot find its location and its neighbor nodes location. Source node will wait until it gets an anchor node in its communication range as shown in Algorithm 1.

Algorithm 1: Gradient-Based Localization

```
If (Source node is GPS enabled node)
  then
Source node sends its location coordinates to all its neighbor nodes
else if (one of the neighbor node is anchor node)
  then
In RREP of anchor node has required information for source node to calculate its location
else
Source node will not transmit the packet until it gets an anchor node in its communication range
```

3.4 Relay Node Selection

After finding the location of all neighbor nodes, the source node stores all the required information in its neighbor list table. Relay node is selected from one of the neighbor nodes for packet transmission. The Cartesian plane is drawn as sink as origin (0, 0), so sink location is known. Knowing the location of neighbor nodes and keeping source node as reference in network Cartesian plane, and selecting a node which is near to the sink as relay node for better performance.

4 Simulations and Results

4.1 Simulation Parameters

Simulation is done in discrete event simulator OMNeT++, for studying wired and wireless network protocols. Simulation is done in terrain size of 200 * 200 m^2, whereas network size varies from 10, 20, 30, 40, and 50 apart from this network that has 1-sink node. Mobility of nodes follows random waypoint mobility for nodes, and as for sink it is no mobility. Keeping sink as origin Cartesian plane is generated in canvas. These nodes use linear battery model to analyze the performance of gradient-based approach and gradient-based localization protocol. Performance metric like throughput, Packet Delivery Ratio, and End-to-End Delay is considered.

4.2 Throughput

It is observed from Fig. 4 that if number of nodes increase there is increase in throughput. However, in 40 numbers of nodes, the throughput is increasing to larger level in gradient-based localization. Because when there is increase in number of nodes, it has better chance to meet the other nodes.

4.3 End-to-End Delay

End-to-End Delay is shown in Fig. 5, where in gradient-based localization when number of node increases delay decreases due to frequent transmission of packet to sink with the help of relay nodes. As in traditional gradient-based routing, when number of nodes increase packet will be drifting within the network, and probability of packet reaching the sink is very low.

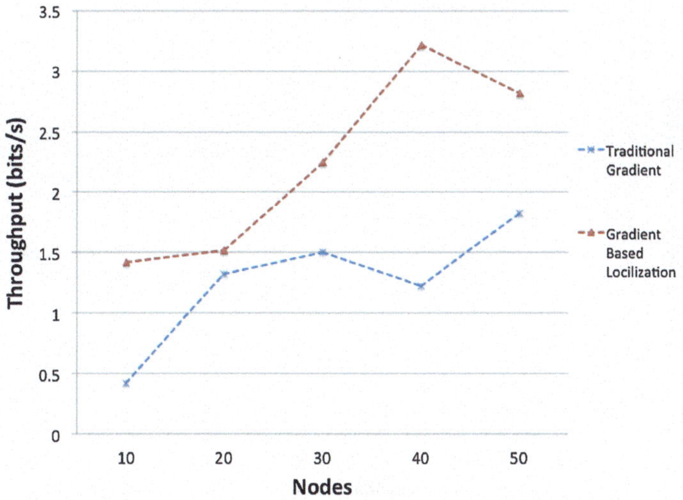

Fig. 4 Throughput

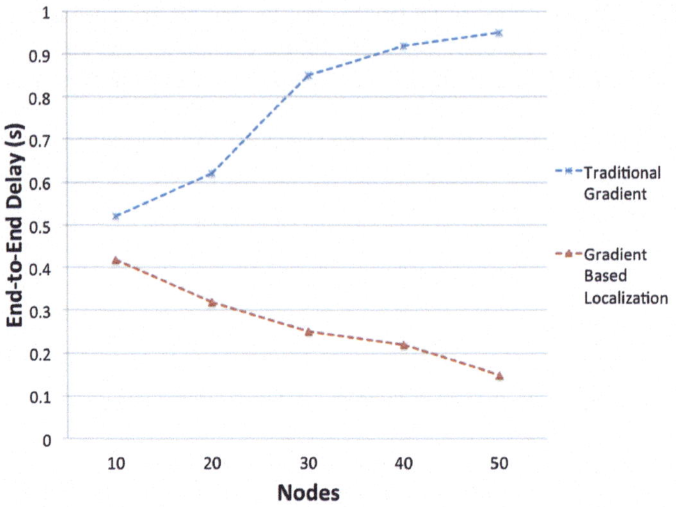

Fig. 5 End-to-end delay

4.4 Packet Delivery Ratio

PDR is displayed in Fig. 6, and here when number of nodes increase there is increase in Packet Delivery Ratio. With the help of localization algorithm, we can find the best relay node to transmit the packet so it reaches the sink node. In existing routing protocol, the packet is transmitting within node network.

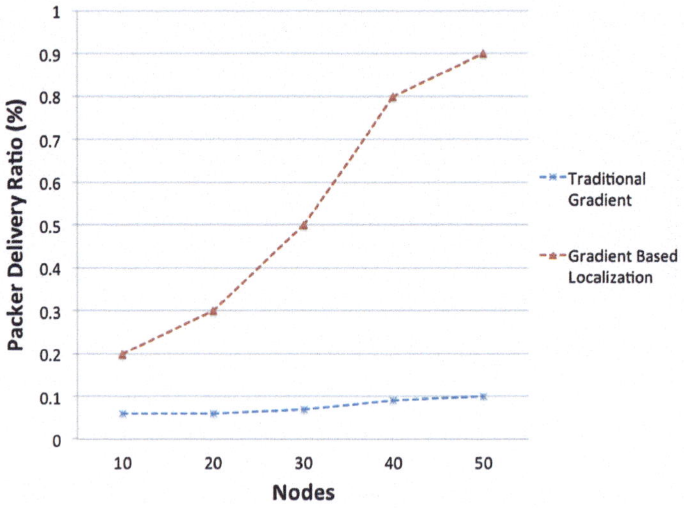

Fig. 6 Packet delivery ratio

5 Conclusion

In this paper, the efficient relay node selection process and gradient-based localization routing protocol for emergency rescue DTN are presented. A Gradient based routing protocol to choose an efficient relay node using localization technique to transmit the packet to reach the sink is simulated. Based on the network Cartesian coordinates and local Cartesian plane value location of the neighbor node is determined. By choosing the best relay node, which is near to the sink, data can be transmitted to the destination during the emergency rescue scenario. The proposed work shows the improvement in the performance metric like Packet Delivery Ratio, throughput, and End-to-End Delay for smaller number of nodes.

To improve the performance of proposed gradient-based localization routing protocol, mobility prediction technique is used to select the relay node along with localization technique. Also, we can work on the network with less overhead and implement for larger area network is considered in future.

Acknowledgements The authors would like to thank the anonymous reviewers for their comment in improving the paper and also we extend our gratitude to VIT University, Chennai, for their support.

References

1. Patil AP, Rajani Kanth K, BatheySharanya, Dinesh Kumar MP, Malavika J (2011) Design of an energy efficient routing protocol for MANETs based on AODV. IJCSI Int J Comput Sci Issues 8(4) (No 1)
2. Hoebeke J, Moerman I, Dhoedt B, Demeester P (2004) An overview of mobile ad hoc networks: applications and challenges. J-Commun Netw 3(3):60–66
3. Koushik CP, Vetrivelan P, Ratheesh R (2015) Energy efficient landmark selection for group mobility model in MANET. Indian J Sci Technol 8(26)
4. Wang Y, Lin F, Wu H (2005) Efficient data transmission in delay fault tolerant mobile sensor networks (DFT-MSN). In: Proceeding of IEEE international conference on network protocols, Nov 2005
5. Kuriakose J, Joshi S, George VI (2014) Localization in wireless sensor networks: a survey. In: International conference on information communication and embedded system (ICICE2014)
6. Spano D, Ricciato F (2016) Opportunistic time-of-arrival localization in fully asynchronous wireless networks. Pervasive Mob Comput
7. Das AP, Thampi SM (2016) Fault-resilient localization for underwater sensor networks. Ad Hoc Netw
8. Zhang Q, Sobelman GE, He T (2009) Gradient-based target localization in robotic sensor networks. Pervasive Mob Comput 5(1):37–48
9. Boukerche A, Oliveira HABF, Nakamura EF, Loureiro AAF (2007) Localization systems for wireless sensor networks. IEEE Wirel Commun 14(6):6–12
10. Baudic G, Auger A, Ramiro V, Lochin E (2016) Using emulation to validate applications on opportunistic networks. arXiv:1606.06925

11. Auger A, Baudic G, Ramiro V, Lochin E (2016) Demo: using the HINT network emulator to develop opportunistic applications. In: Proceedings of the eleventh ACM workshop on challenged networks, pp 35–36
12. Kanai H, Koizumi Y, Ohsaki H, Imase M (2012) Gradient-based routing in delay tolerant mobile sensor networks incorporating node mobility. In: 2012 IEEE consumer communications and networking conference (CCNC). IEEE, pp 235–239

Modelling and Performance Analysis of Wi-fi Offloading

Liji A. Jose and C. Hemanth

Abstract Mobile data offloading or Wi-Fi offloading is one such approach to influence the unused bandwidth across different wireless technologies where data flowing through cellular network is transferred. Here the delayed type of offloading is considered for modelling. The traffic in delayed offload would be traffic on per content basis has loose QoS guarantees that's were individual packets can be delayed but entire packet should reach on time and also truly delay tolerant traffic. These packets can be introduced with an appropriate queuing mechanism such as reneging, balking or a probabilistic method and modelled for analysis of different metrics. In this paper, the performance analysis for delayed type of mobile offloading in access point-based method for the metric of efficiency is modelled and analysed. In order to verify the optimality of proposed model, the packet arrival rate and average delays are estimated through simulation using MATLAB software tool.

Keywords Wi-fi offloading · Delayed network · Efficiency · QoS

1 Introduction

In the modern era when the data traffic is having an exponential increase, the greed to provide a promising and low-cost solution to manage the data traffic increases. Also, the investments for access networks, infrastructure and the licensed spectrum availability pose a vulnerable need for alternative methods to reduce the pressure on cellular networks. Few methods were initiated by operators to reduce the contingency over mobile networks.

L. A. Jose · C. Hemanth (✉)
School of Electronics Engineering (SENSE), VIT University,
Chennai Campus, Chennai, India
e-mail: hemanth.c@vit.ac.in

L. A. Jose
e-mail: liji.jose49@gmail.com

A. M. Zungeru et al. (eds.), *Wireless Communication Networks and Internet of Things*, Lecture Notes in Electrical Engineering 493,
https://doi.org/10.1007/978-981-10-8663-2_4

Policies such as choking connection speed and capping data usage, which failed in order to provide customer satisfaction, thus a broader approach was introduced called as mobile data offloading. Reliability of traffic delivery over time constraints leads offloading to delayed and non-delayed types, further on the type of their deployment as infrastructure, access point based and terminal to terminal based. Issues to be addressed in non-delayed types include handover transparency and between existing cellular of past and other alternative access technologies share interoperation. An intuitive approach is to leverage the unused bandwidth across different wireless technologies.

Mobile data offloading is considered as the use of a complementary wireless technology to transfer data originally targeted to flow through the cellular network, in order to improve some key performance indicators. Simply when an accessible Wi-fi connection is available, the sessions will be transferred to Wi-fi saving cost and energy (during low-throughput data transfer) to the user. The Wi-fi offloading is classified as infrastructure based, access point (AP) based, terminal to terminal (T2T) based, based on the type of deployment and as delayed offloading, non-delayed offloading based on the type of delay.

2 Literature Review

Over the years, a lot of work is proposed and carried out in the field of data offloading considering few prominent works. Donguen Suh through "Efficiency analysis of Wi-Fi Offloading Techniques" proposed consideration of two types of Wi-fi offloading techniques entitled as opportunistic Wi-fi offloading, where opportunistically meeting Wi-fi access points (APs) and mobile node tend to be the only condition for data offloading. Also delayed Wi-fi offloading, where with the expectation of future AP contacts data transfer is delayed. They also formulated analytical models on Wi-fi offloading efficiency as the ratio of the amount of offloaded data to the total amount of data [1]. Insook Kim through "Probabilistic Offload Scheme in Integrated Cellular WiFi Systems" proposed Wi-fi offloading problem in an integrated cellular Wi-fi system consisting of mobile base station (MBS) and Wi-fi access point (AP). They propose a probabilistic offloading scheme, where cellular packets that arrive at the queue of MBS are offloaded to the queue of Wi-fi AP with an offload probability. The offload probability is determined to minimize the average delay experienced by the cellular packets while guaranteeing stability of both cellular and Wi-fi system. Studied two priority disciplines: First-In-First-Out (FIFO) and Non-Preemptive Priority Rule (NPPR). Considered the congestion in network, penalty switching and pricing of network in congestion-aware network selection problem in data offloading. The offload probability p is determined to minimize the average delay experienced by cellular packets while guaranteeing stability of both systems (cellular and Wi-fi) [2].

Kyunghan Lee through "Mobile Data Offloading: How Much Can WiFi Deliver?" proposed study on the 3G performance through Wi-fi networks of mobile data offloading. They recruited about 100 iPhone users from metropolitan areas and collected statistics on their Wi-fi connectivity during about a two and half week period. Acquired simulation traces indicate that Wi-fi already offloads about 65% of the total mobile data traffic and save 55% of battery power without using any delayed transmission. In case the data transfers are delayed using a deadline till users enter a Wi-fi zone, achievement of substantial gains when a fairly larger deadline of tens of minutes [3]. Joohyun Lee through "Economics of WiFi Offloading: Trading Delay for Cellular Capacity" proposed, how much economic benefits can be generated due to delayed Wi-fi offloading, based on a two stage sequential game the interaction between users and a single provider and users is modelled. They analytically first proved that Wi-fi offloading for both the provider and users is economically beneficial. Also to quantify the practical gain, a trace-driven numerical analysis is conducted. The revenue gain from the delayed offloading generated by network upgrade from 3G to 4G is similar to the on-the-spot delayed offloading. The revenue increase of complex pricing schemes (such as two-tier from flat and congestion from volume) becomes smaller for higher offloading chances, which is true as of now and in the future, when more Wi-fi APs are expected to be deployed [4].

Qimei Chen through "Rethinking Mobile Data Offloading in LTE and WiFi Coexisting Systems" proposed to transfer Wi-fi users to the LTE-U network and simultaneously allocate some unlicensed spectrum to LTE-U. In this way, a win–win situation could be generated since LTE can achieve better spectrum efficiency than Wi-fi in the unlicensed spectrum. They utilize the Nash bargaining solution to design fair unlicensed spectrum allocation between Wi-fi and LTE-U, and thereby, a win–win strategy is developed, whose performance is demonstrated by numerical simulation. Through numerical simulation, compared the three different user transfer schemes based on the availability of CSI as random transfer, distance based transfer and CSI based transfer [5]. Eyuphan Bulut through "WiFi Access Point Deployment for Efficient Mobile Data Offloading" proposed the deployment of Wi-fi access points (APs) in a metropolitan area for efficient offloading of mobile data traffic. They proposed a deployment algorithm by analysing a large-scale real-user mobility trace based on the density of user data request frequency. Proposed to deploy the APs to the locations with the highest density of user data access requests. Ray-Guang Cheng et al., "Offloading Multiple Mobile Data Contents Through Opportunistic Device-to-Device Communications" presented a popularity-based relaying user selection algorithm to determine the number of relaying users for distributing multiple contents with different popularity. An analytical model is then presented to estimate the amount of reduced mobile data traffic under single-hop and multi-hop opportunistic forwarding scenarios [6].

Fidan Mehmeti through "Performance Analysis of Mobile Data Offloading in Heterogeneous Networks" proposed a queueing analytic model that can be used to understand the performance improvements achievable by Wi-fi-based data offloading, as a function of Wi-fi availability and performance, user mobility and traffic load, and the coverage ratio and respective rates of different cellular technologies available. They dealt with: (i) on-the-spot offloading (ii) provided closed-form results and approximations (iii) provided an extension for generic packet size distributions (iv) validated their theory against realistic parameter values and distributions (v) provided some insight about the offloading gains that are of interest to both users and operators [6].

A. Rajabi through "Flows with Bounded Waiting Time in Networked and Distributed Systems" studied the performance of queueing systems in which customers arrive with a probability balk, and with a deterministic distribution renege. The probabilities of the number of customers for balking system have been derived. An intuitive relation between the average number of customers in balking and reneging systems is presented. Finally, a straightforward mechanism to calculate the jitter and average degree of multiplexing has been provided. Such mechanism is also useful in analysis of fair queueing and weighted fair queueing which are recently employed in Internet routers to provide a degree of quality of service (QoS) among different traffics [7].

3 System Specification

Consider a mobile system with input arrival "Lm" and output transfer rate as "Um" also consider a Wi-fi system with input arrival "Lw" and output transfer rate of "Uw". The mobile system buffer is under constant pressure because of minimal available bandwidth constraint the mobile node is always under scrutiny to transfer data whenever it comes in contact with a Wi-fi system which shares the idea of offloading. In the specified system model, the system is considered to be a delayed system such that whenever there is a congestion in the mobile system the mobile system will transfer its data to the Wi-fi network with a proportionate amount of delay. The offloading mechanism is achieved based on a probabilistic condition such that a threshold probability is set at "p", whenever the congestion in the mobile system "x" is greater than the threshold probability the offload occurs (Fig. 1).

Fig. 1 System model

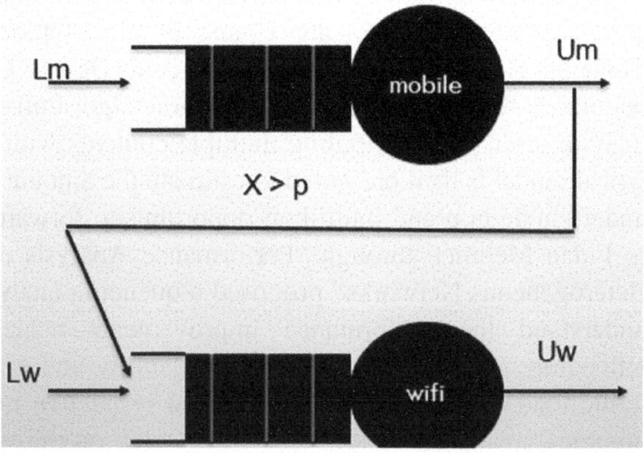

4 Results

The analysis for modelling and performance of Wi-fi offloading are performed in MATLAB; the analysis and s for the same are discussed in Figs. 2, 3, 4 and 5.

```
Command Window
New to MATLAB? Watch this Video, see Examples, or read Getting Started.
>> WiFi
Current plot held
Current plot held

Times =

    4.1089  107.7676   53.5686  149.6949   28.5994

User 1 arrives into the system at time 4.1089
User 1 leaves the system at time 37.9907
User 2 arrives into the system at time 111.8765
User 2 leaves the system at time 138.1611
User 3 arrives into the system at time 165.445
User 3 leaves the system at time 192.0611
User 4 arrives into the system at time 315.14
User 5 arrives into the system at time 343.7394
User 4 leaves the system at time 17748.3657
User 5 leaves the system at time 17764.1758
```

Fig. 2 Initial time specifications

```
>> WiFi
Current plot held
Current plot held

Times =

   20.1745   14.0764  247.1763   91.6834  134.7572

User 1 arrives into the system at time 20.1745
User 2 arrives into the system at time 34.2509
User 2 leaves the system at time 172.0535
User 1 leaves the system at time 197.3805
User 3 arrives into the system at time 281.4272
User 3 leaves the system at time 315.7034
User 4 arrives into the system at time 373.1106
User 4 leaves the system at time 409.0632
User 5 arrives into the system at time 507.8678
User 5 leaves the system at time 546.719
>>
```

Fig. 3 Further time specification

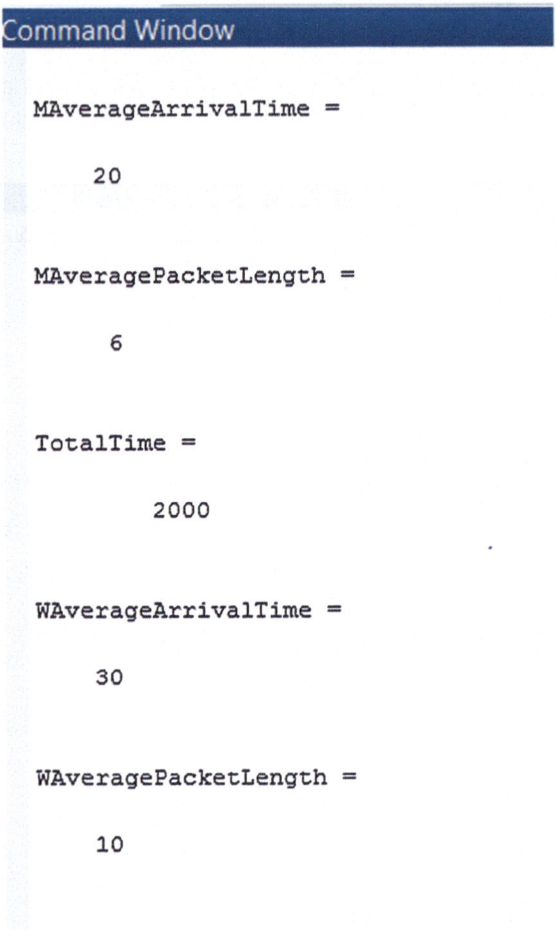

Fig. 4 Wi-fi and cellular packet arrival rate and lengths

Initially, the calculations for initial arrival and further packet arrivals are calculated based on which the probabilistic offload is done. The packet arrivals at the initial time when the user approaches to the access point to the time intervals at the final when offload is done are estimated. Based on the estimated values, the probabilistic offload condition is specified and offloading is done. Lm and Lw are poison distribution value for mobile and Wi-fi, respectively.

This further contributes to the estimation of probability of offload based on which the efficiency is estimated.

```
WAveragePacketLength =

    10


TotalTime =

        2000


Lm =

    0.9911


Lw =

    0.9906

probab offload

A =

3/2 - 2000/(2019*((1000*p)/1009 - 2057/4038)^2) - 1/((1000*p)/1009 + 9/1009)^2
```

Fig. 5 Offload probability

References

1. Suh D et al (2016) Efficiency analysis of Wi-Fi offloading techniques. IEEE Trans Veh Technol 65(5):3813–3817
2. Kim I et al (2015) Probabilistic offload scheme in integrated cellular WiFi systems. In: 2015 9th international conference on next generation mobile applications, services and technologies
3. Lee K et al (2010) Mobile data offloading: how much can WiFi Deliver? In: ACM CoNEXT 2010, 30 Nov–3 Dec 2010, Philadelphia, USA
4. Lee J et al (2014) Economics of WiFi offloading: trading delay for cellular capacity. IEEE Trans Wirel Commun 13(3):1540–1554
5. Chen Q et al (2016) Rethinking mobile data offloading in LTE and WiFi coexisting systems. In: IEEE wireless communications and networking conference (WCNC 2016)—Track 2—MAC and cross layer design
6. Bulut E et al (2012) WiFi access point deployment for efficient mobile data offloading. In: Proceedings of ACM workshop Pingen at Mobicom'12, Turkey, Aug 2012
7. Cheng R-G et al (2015) Offloading multiple mobile data contents through opportunistic device-to-device communications. Wireless Pers Commun 84:1963–1979. https://doi.org/10.1007/s11277-015-2492-1

Integrity Verification for Shared Data in Group with User Revocation

M. Suguna, S. Mercy Shalinie and R. Sivaranjani

Abstract Cloud computing provides storage for the multiple users to store and share their data anywhere at anytime basis. There were some security issues faced by the cloud users such as data correctness, data theft, data leakage, privacy on user level because of the third-party data control. One of the major issues in cloud storage is ensuring data integrity when data are shared by multiple users in the cloud and the data owner accesses data locally. To overcome this issue, many public integrity auditing schemes have been proposed where the computation overhead is huge for the data owner. Hence, efficient auditing with minimum overhead at client side is in need. In the proposed method, we have multi-user modification model with user revocation where the auditing work is delegated to a trusted third-party auditor (TPA) on a secure model, thereby reducing the overhead faced by client.

Keywords Public auditing · Cloud storage · Third-party auditor
Block less verification · User revocation · Data integrity

1 Introduction

Cloud computing provides storage for the users to access the data on their own computer's. Cloud is used to connect multiple computers via the digital network through one computer. Some cloud memory such as cloud-based software Dropbox [1] constructs the cloud application. CloudMe [2] has been built as a cloud application. There are two components in cloud architecture; they are front end and back end. The front end is only accessed by client or user, and back end is full of cloud architecture; here cloud controls the storage devices and servers. Cloud storage is a model to store data on multiple virtual servers hosted by TPA rather

M. Suguna (✉) · S. Mercy Shalinie · R. Sivaranjani
Department of Computer Science and Engineering,
Thiagarajar College of Engineering (TCE), Madurai, India
e-mail: mscse@tce.edu

A. M. Zungeru et al. (eds.), *Wireless Communication Networks and Internet of Things*, Lecture Notes in Electrical Engineering 493,
https://doi.org/10.1007/978-981-10-8663-2_5

than being hosted on dedicated servers. There are some types of cloud for user flexibility; they are public cloud, private cloud, community cloud, hybrid cloud.

Cloud computing are of three types: Infrastructure as a Service (IaaS): The IaaS is the base for the cloud. By using the IaaS, the CSP can ensure that the data are secure and the data can be accessed via, firewalls, routers, storage, and other network equipment in the cloud; Platform as a Service (PaaS): In PaaS, a client can create own appliance which runs on contributor infrastructure; and Software as a Service (SaaS): In SaaS, there is no requirement of client side expenditure for servers or software licenses. There are some of the security issues faced by cloud computing which are data integrity, data theft, security on vendor and user level, information and physical security, third-party data control, operational security. Two kinds of threats are prevalent in shared data storage in cloud. First, the client can try to corrupt the data in the shared pool. Second, the CSP can accidently remove or change the data in its memory due to hardware/software crash. The major issue in cloud storage is data integrity. To solve the problem, many mechanisms [3–6] have been proposed and allowed multiple users to conduct integrity checking beyond downloading the whole data from the cloud. In existing, data owner who carry the secret key can only change the data and share in the cloud. To allow group user modification with integrity [7], the data owner needs to stay online, collect the changes made on data from the other clients, and update the verification tags [7] for each modified user with integrity assurance.

In cloud, to support multi-user modification, Wang proposes data integrity upon ring signature [5]. In this scheme, auditing cost is maintained with fixed size in the group. The cloud node is responsible for updating signature in the cloud storage to prevent impersonation attack in the cloud [5]. We need to overcome the following challenges to get efficient user revocation:

(1) Allowing group user to modify or share information in the cloud without the help of data owner and creating individual aggregate tags for each user becomes a problem. This is because the authentication tags must be generated with client's secret key, which is kept secret from all. Without verification tags, user cannot provide integrity verification in cloud. To solve the problem, let users can share the same secret key. By this, all verification tags are in the similar format, and it can be easily coagulated.
(2) Efficient user revocation. All users authenticated tags are needed to be updated in the cloud, and all revoked users authentication tags are also updated and maintained in the cloud so that we can easily remove the secret key of revoked user from the cloud. If any user revoked from the group, then public key of the group is needed to be updated with authentication tag in the cloud.
(3) Public integrity verification. Public auditing is handled by the data owner and also by any clients who hold a public key. In this scheme, we propose a novel integrity auditing scheme for cloud environment to support multiple-user modification which addresses the above challenges. This scheme supports polynomial-based verification tags from multiple clients into one and transfers the information to the auditor. In this scheme, auditing cost is maintained with

fixed size in the group to support group user revocation [4, 8]. The cloud node is responsible for updating the signature in the cloud storage to prevent impersonation attack in the cloud [7, 9]. By using Shamir's Secret Sharing [10], secret divides into *N* polynomial shares. The design of public integrity auditing scheme supports group user modification with blockless verification.

2 Models

2.1 System Model

In system model, cloud consists of three systems: cloud server, public verifier (TPA), and group users. Cloud server maintains storage services to the group users. Group user consists of number of clients, and original user shares data in the cloud. All clients in the group can change and access the data in the cloud. TPA can check the integrity of data using proof information from the cloud. Once user revoked, then user cannot access the information in the cloud. As our proposed scheme allows public integrity auditing, any user who holds public key can act as a TPA in the cloud. The acquired information are stored as structure of files and each file splits into number of blocks with the authentication key that is created by the own user. When client modifies or updates the block, client updates the corresponding verification tag with his/her own secret key without contacting the user. If any user revoked from group, the user cannot access the data in the cloud because TPA verifies and recomputes the public key (for more detail refer Fig. 1) for the group users.

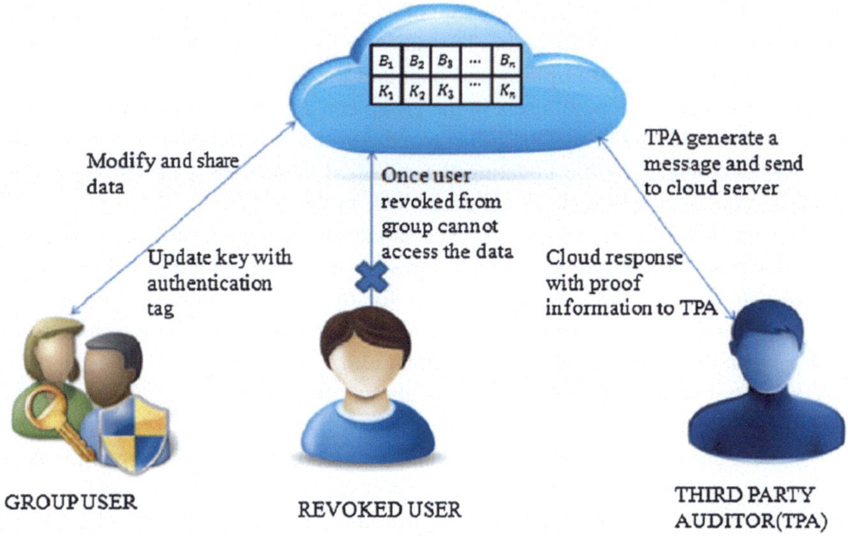

Fig. 1 System model

2.2 *Threat Model*

In threat model, integrity can be disputed in the following ways: cloud service provider (CSP) can also crash the data, hardware or software failure and operational errors of system administrator, revoked users can also try to access the information stored in the cloud.

We analyze the problem of constructing a public integrity auditing for dynamic data shared in a group with user revocation.

(1) Public auditing: The TPA verifies the data block stored in the cloud without downloading the information from the cloud.
(2) Data correctness: The TPA checks the integrity of data shared in the cloud.
(3) Unforgeability: Group user can only generate valid key or signature on shared data.
(4) Efficient user revocation: If any user is revoked from the group, then the user key and the signed blocks are taken by the original user. Then revoked user accesses are removed from the cloud.
(5) Scalability: Multiple users shared their data in the cloud publicly, and the public verifier is able to handle the multiple auditing tasks simultaneously in efficient manner.

3 Proposed Methodology

Setup: In setup step, original user runs key generation part and generates the public key (P_k), private key (U_k), secret key (S_k) of each user. In this design, each user has unique secret key for modification. To audit the file in the data block, each user needs authentication tag to upload and maintain the log in a cloud.

Update: In update step, all users in the group can change or modify the data in a cloud. After modifying or updating the data, the user needs to compute the tag with their own secret key. While updating the modified data block with the tag in the cloud, it simultaneously updates the data block in the log file.

Challenge: Third-party auditor (TPA) evaluates the integrity verification of data. TPA audits the log file and generates a message and sends to the cloud server.

Prove: In proving step, the cloud waits for the message from challenge step; then, the cloud generates the proof information. Finally, the cloud responds to third-party auditor (TPA) with proof information.

Verify: By using the proof information, the verifier checks the file integrity and analyzes the data integrity.

User revocation: The original user and the cloud check if any user is revoked from the group; then, the authentication tag generated by revoked user and a secret key of revoked user are removed from the tag. Then original user checks the number of tags modified by the revoked users which becomes a potential burden for the user'. To control the burden from the original user, all tag update operations are handled by cloud because the cloud can support parallel processing. After receiving the message, cloud updates the authentication tag of each block. The verifier and the group users then remove the public information from revoked users. Public verifier (TPA) audits the files last accessed by the revoked users and sends the message to cloud. The cloud checks the message and log file and sends the proof information to the TPA. TPA checks the integrity of the data and analyzes the report as accept or reject.

Step 1. In the initial setup, an original user s_0 evaluates key algorithm and creates the public key (P_{k}), secret key (U_{k}), private key (S_{k}) for every user. Using file processing algorithm, each file F splits into n blocks of data and each block then divides into s elements. For every user tags σ_i are generated for the files to be uploaded and these tags are stored at third-party verifier.

Step 2. In update step, multiple users can share or modify the data in the cloud simultaneously and a new authentication tag σ_i is computed for each modification or updation done by the user. During download, TPA generates a challenge message

$$\delta_i = e\left(\sigma_i, g^{\frac{\varepsilon_0 R}{\varepsilon_k}}\right) e\left(\left(u^{B_i}.g^{\overrightarrow{f\beta_i}^{(\gamma)}}\right), g\right)^{\varepsilon_0 R} \quad \text{and}$$

$$\delta_i e\left(\sigma'_{i,}\left(g^{\frac{\varepsilon_0}{\varepsilon_0 + \tau}}\right)^R\right) e\left(\left(u^{B_i}.g^{\overrightarrow{f\beta_i}^{(\gamma)}}\right), g\right)^{\varepsilon_0 R}$$

sends it to the cloud, and cloud acquires the challenge message from the auditor (TPA) and creates proof information and sends it to the auditor.

Step 3. By utilizing the proof information, verifier checks data correctness verification on download. The original user s_0 runs a Shamir's Secret Sharing scheme and generates N points. Each cloud node needs to update a piece of the tag. If any client is revoked from the group, then the group key is updated and the updated key is circulated amongst all group users.

Step 4. By this, public auditing and user revocation are achieved securely using a trusted verifier. By using dynamic auditing scheme, any client in the group can easily modify and update blocks in the data in single block using dynamic operation.

Multi-file auditing: In cloud, group users often make changes in blocks to ensure data integrity TPA audit the data in blocks frequently. So the computational cost is inefficient. To control batch auditing operation performed in the cloud. To audit N number of information blocks in file batch, challenge converts N number of data blocks into one message and one verification step to reduce cost. By this multi-file, auditing enables the verifier to perform integrity auditing for N number of files as single file cost.

4 Support Dynamic Operations

Any client in the group can easily change the information in the cloud using dynamic operation. Dynamic operation supports insert, delete, update on single block. By using index hash table, all users can efficiently perform dynamic operation on shared data. Client can modify the single data block in shared data by using insert and delete operations. The modified blocks, are all changed and if users share the data, then the signature of the block has been recomputed the signature of the block even though the content has not been changed. Here, I denotes Index and B denotes block in the table.

By using hash table [5], user can perform dynamic operation efficiently. In our appliance, the identifier is described as $id_j = \{V_j, r_j\}$ where v_j is denoted as the virtual values of blocks a_j and r_j is a value created by a hash basis H_2. The value of r is generated by the H_2; it shows that each block has a solitary identifier and the virtual indices are able to ensure that all shared data are in right order in index table. (Figs. 2 and 3 show the multiple dynamic operations with our index hash table). Here, ρ supports sufficient number of blocks for the group, so that there is no way to have the same virtual indexes in the table.

I	B	V	R
1	a_1	ρ	r_1
2	a_2	2ρ	r_2
3	a_3	3	r_3
4	a_4	4ρ	r_4
⋮	⋮	⋮	⋮
N	a_n	$n\rho$	r_n

Insert →

I	B	V	R
1	a_1	ρ	r_1
2	a_2'	$\|3\rho/2\|$	r_2'
3	a_2	2ρ	r_2
4	a_3	3	r_3
5	a_4	4ρ	r_4
⋮	⋮	⋮	⋮
n	a_n	$n\rho$	r_n

Fig. 2 Insert block into dynamic data operation using hash table as identifier

I	B	V	R
1	a_1	ρ	r_1
2	a_2	2ρ	r_2
3	a_3	3	r_3
4	a_4	4ρ	r_4
⋮	⋮	⋮	⋮
N	a_n	$n\rho$	r_n

Update →
Delete →

I	B	V	R
1	a_1'	ρ	r_1'
2	a_2	2ρ	r_2
3	a_3	3	r_3
4	a_4	4ρ	r_4
⋮	⋮	⋮	⋮
n	a_n	$n\rho$	r_n

Fig. 3 Update blocks and delete blocks in dynamic data operation using a hash table as identifier

5 Performance Analysis

In this mechanism, we evaluate the performance of the proposed method by storing the files on CloudMe and implement the algorithms using Java. On CloudMe, we deploy different text files accessed by the users and modify file with the authentication tag. The computational cost is calculated for the user and verifier by varying the file size. The communication cost is analyzed through the challenge message and proof information. To check the verification tag generation time, we increase the number of blocks in the file. Our result depicts the analysis of verification tag generation. To verify the file size in the auditing, we alter the number of blocks from 1000 to 100,000. As depicted in Fig. 4, the tag generation time is propositional to the number of blocks from 10 to 100 s.

Figure 5 shows that to revoke a user, the advanced user revocation algorithm consumes minimal storage overhead for tag updation for each user which leads to increase in communication cost.

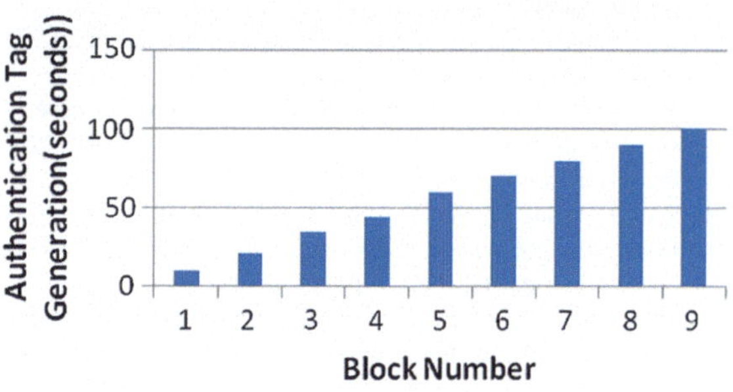

Fig. 4 Authentication tag generation time

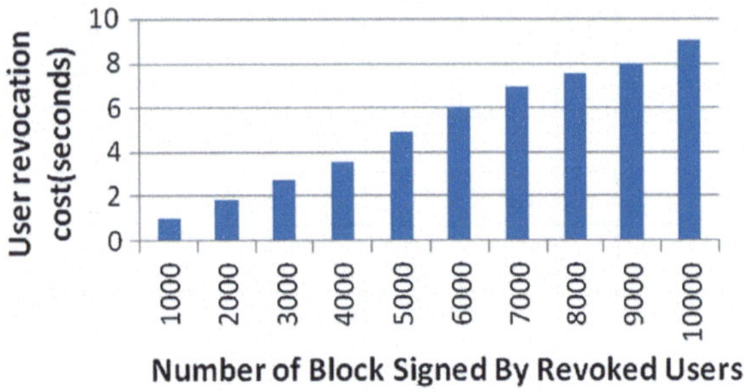

Fig. 5 User revocation cost on cloud

6 Conclusion

Public integrity auditing mechanism checks the data correctness in cloud-sharing resources. To support group user modification and dynamic auditing, user generates an authentication tag to insert, delete, and update the data in the block. TPA can verify the data integrity with blockless verification. Authentication tag generation is performed by user revocation algorithm. Although the advanced user revocation algorithm requires more cost and cloud-sharing resources, it achieves better reliability for the system. In this scheme, we extend our mechanism to support batch auditing but there are some issues that will be continued as a future work. One of them is traceability, which means ability to reveal the identity of the signer based on verification meta data. Another issue is the cloud reciprocity problem (although original user back up his/her data in multiple CSPs, CSPs might exercise mutual aid to avoid the huge cost of data lost). Thus, we can achieve data correctness for multiple tasks through batch auditing technique.

References

1. Dropbox (2007) A file-storage and sharing service. Dropbox [Online]. Available: http://www.dropbox.com/
2. CloudMe. A file-storage and sharing service in cloud. CloudMe [Online]. Available: http://www.CloudMe.com/
3. Wang C, Wang Q, Ren K, Lou W (2010) Privacy-preserving public auditing for data storage security in cloud computing. In: Proceedings of the 29th IEEE international conference on computer communications (INFOCOM), San Diego, CA, USA, Mar 2010, pp 1–9
4. Zhu Y, Wang H, Hu Z, Ahn GJ, Hu H, Yau SS (2011) Dynamic audit services for integrity verification of outsourced storages in clouds. In: Proceedings of the ACM symposium on applied computing (SAC), pp 1550–1557
5. Wang B, Li B, Li H (2012) Oruta: privacy-preserving public auditing for shared data in the cloud. In: Proceedings of the IEEE 5th international conference on cloud computing (CLOUD), Washington, DC, USA, Jun 2012, pp 295–302
6. Jiang T, Chen X, Ma J (2016) Public integrity auditing for shared dynamic cloud data with group user revocation. IEEE Trans on Comput. Citation information: https://doi.org/10.1109/tc.2015.2389955

7. Yuan J, Yu S (2015) Public integrity auditing for dynamic data sharing with multiuser modification. IEEE Trans Inf Forensics Secur 10(8):1717
8. Wang B, Li B, Li H (2013) Public auditing for shared data with efficient user revocation in the cloud. In: Proceedings of the 32nd IEEE international conference on computer communications (INFOCOM), Turin, Italy, Apr 2013, pp 2904–2912
9. Wang B, Li B, Li H (2015) Panda: public auditing for shared data with efficient user revocation in the cloud. IEEE Trans Serv Comput 8(1)
10. Shamir A (1979) How to share a secret. Commun ACM 22(11):612–613

Shortest Path Solution to Wireless Sensor Networks Using Edge-Based Three-Point Steiner Tree Concept

S. Sundar, V. Balakrishnan, R. Kumar and Harish M. Kittur

Abstract Wireless sensor networks have gained immense significance and popularity in modern technology especially as a result of emerging concepts like the Internet of Things (IoT). These networks are usually under constant pressure to scale-up in order to meet the ever-growing demand. Under such situations, it is often required to connect two existing adjacent, but independent, networks so that they form a single larger network. Some of the networks may even have mobile nodes. This paper proposes an effective method for placing a bridge node in the intersection of coverage region of two networks using a combination of Steiner tree algorithm and concept of edge nodes—Edge-based Three-Point Steiner (EdTPS) tree—such that communication across the network can be carried out with the shortest possible path.

Keywords Steiner tree · Bridge node · Mobile node · Edge-based three-point Steiner tree

1 Introduction

The amount of traffic within networks always shows an upward trend. This occurs not only due to traffic generated by nodes already present but also due to the need for expanding the coverage. In order to meet these demands, separate networks are

S. Sundar (✉) · V. Balakrishnan · H. M. Kittur
School of Electronics Engineering, VIT University, Vellore 632014, Tamil Nadu, India
e-mail: sundar.s@vit.ac.in

V. Balakrishnan
e-mail: bvs281994@gmail.com

H. M. Kittur
e-mail: kittur@vit.ac.in

R. Kumar
WIPRO Technologies, Chennai, Tamil Nadu, India
e-mail: rajagopal.kumar@wipro.com

A. M. Zungeru et al. (eds.), *Wireless Communication Networks and Internet of Things*, Lecture Notes in Electrical Engineering 493,
https://doi.org/10.1007/978-981-10-8663-2_6

usually linked to one another. This process of linking presents a lot of challenges. The linking path may have several spatial constraints. Moreover, it is usually more expensive. Hence, the objective is to achieve the link with as few new nodes as possible. In the most constrained scenario, it may only be possible to place one new node to connect the two networks. This one node can be made mobile if the component networks have mobile nodes. In this case, the optimum placement of the node becomes very crucial. The newly placed node has to handle all of the inter-network traffic. Thus, every such communication must happen as fast as possible. This limit is achieved when there exists shortest path communication between any two nodes across the network.

While many algorithms exist which find the shortest path among a given set of points, very few deal with the aspect of placing a new point such that the network as a whole is more efficient. The Minimum Steiner Tree (MST) problem is one such approach. It aims at finding the shortest path interconnecting a set of N points by adding $N - 2$ new points of interconnection. However, in the problem scenario, there are constraints over the number and location of the additional nodes. Hence, the pure Steiner tree approach may not yield best results as we are required to place only one new node.

As mentioned above, if only one new node is to be placed, we may apply Steiner tree algorithm to various combinations of sets of 3 points ($N = 3$) and find the best combination so that we get only one solution point. This is the brute force approach. However, this approach might also yield sub-optimal results for certain arrangement of nodes as demonstrated in this paper. The proposed method addresses the above-mentioned drawbacks and yields better results by using the concept of edge node in conjunction with the three-node Steiner tree problem.

2 Steiner Tree Problem and Related Work

For the fixed set of V vertices, the Euclidean Steiner tree problem is to find the tree with minimum Euclidean length spanning all vertices in V, while allowing for the addition of extra or auxiliary vertices called as Steiner vertices. This is an NP-hard problem [1], and hence, it is very difficult to obtain polynomial time algorithms for exactly solving it.

This section briefs about few Steiner tree-based works which are based on approximate algorithms where the detailed survey is found in [2].

In 1993, a simple greedy algorithm was introduced by Zelikovsky using the concept of 3 Steiner trees. This method has an approximation ratio of 1.834 [3]. Later, Berman and Ramaiyer have extended the approach to k-tree with an approximation ratio of 1.734 [4]. Karpinski and Zelikovsky proposed the concept of loss of a Steiner tree which is defined as the cost of the minimum spanning forest in the Steiner tree [5]. An approximation ratio of 1.644 was achieved with an algorithm that minimizes the loss of a Steiner tree and sum of edge cost. All approximation algorithms proposed by the researchers for the Steiner tree problem have borrowed from Zelikovsky's concept [6].

The problem of Steiner tree is quite similar to the minimum spanning tree problem [6]. The primary difference is that in Steiner tree problem, extra edges and vertices may be introduced to minimize the length of the spanning tree. The deriving of minimum cost tree is quite useful for several applications, namely VLSI physical design [7], telecommunication network design [8], multicast packing [9], [10], network topology control [11, 12], maximizing lifetime of networks [1] and network design [6].

As the Steiner tree problem is of having great significance for both theoretical research and practical applications, we have proposed optimum placement of a Steiner node to extend the coverage and allowing for establishing shortest path between two networks.

3 Methodology

The entire work is simulated using MATLAB, a numerical computing environment with multiple paradigms and fourth-generation programming language. It also provides good graphical support for visualization of the networks.

The workspace is represented by a 100 by 100 graph spanning along the horizontal (x)-axis and vertical (y)-axis. The left half represents one network while the right half represents the second network. The region of common network coverage is assumed to be in the middle of the two networks. For simplicity, this common region may be represented by the line $x = 50$. Any number of nodes may be placed on either side. These would represent the individual networks. The objective is to find the location in the common region ($x = 50$) where the bridge node is to be placed such that there exists shortest path communication between the two networks (Fig. 1).

The placement of individual nodes can be done by clicking the mouse at the corresponding points on the graph. The inbuilt commands of MATLAB provide all necessary graphical input and output support for the purpose of simulation (Figs. 2, 3, 4 and 5).

Once the positions of nodes of each network is fixed, the Edge-based Three-Point Steiner (EdTPS) tree algorithm is applied,

1. Find the set of edge nodes for network 1 and 2.
2. Based on the total number of edge of nodes (N1edge + N2edge), there may be two cases.

   Case (N1edge + N2edge) < 2: Find the rightmost node of network 1 and leftmost node of network 2. The solution Bfinal will be the mid-point of these two nodes.

   Case 2 (N1edge + N2edge) > 2: Apply Steiner Tree algorithm to all edge nodes three at a time and find the sum of paths.

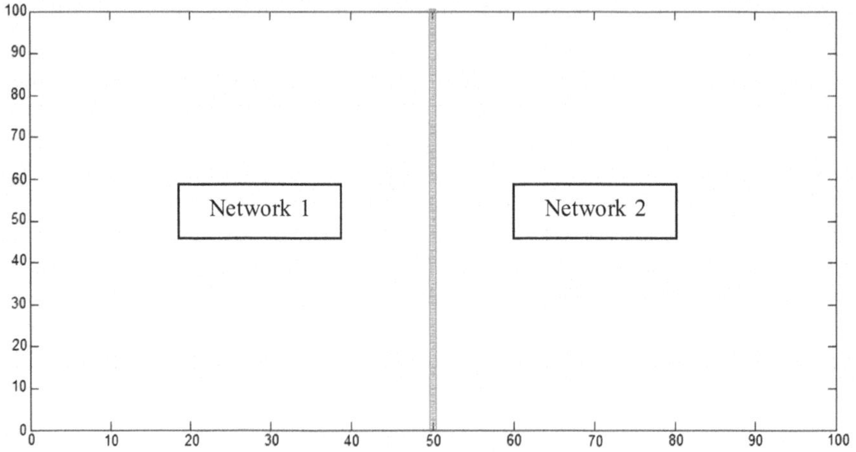

Fig. 1 Layout of networks

Fig. 2 Nodes placed by simply clicking the mouse

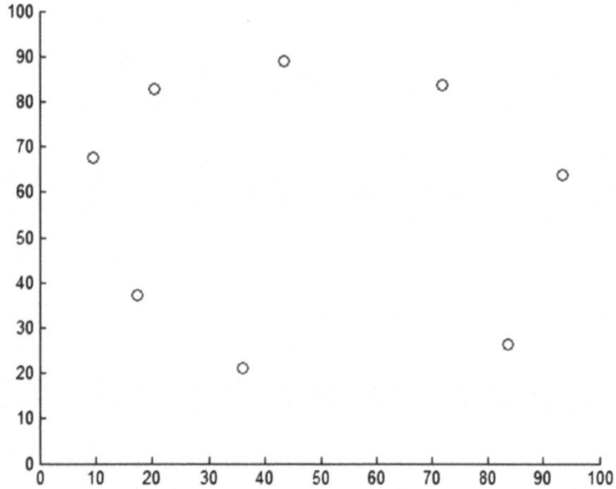

3. Find the set of three nodes (with at least one node from each network) which have the least sum of paths.
4. Obtain the Steiner point for these three points. If the Steiner point is not on the line $x = 50$, extrapolate the line joining the Steiner point and the node in that network which has only one of the three nodes mentioned in step III, until it intersects the line $x = 50$.
5. This newly obtained point is the required optimum placement of the bridge node.
6. Such points can be obtained at regular intervals to implement a network with mobile nodes like mobile ad hoc networks (MANETs).

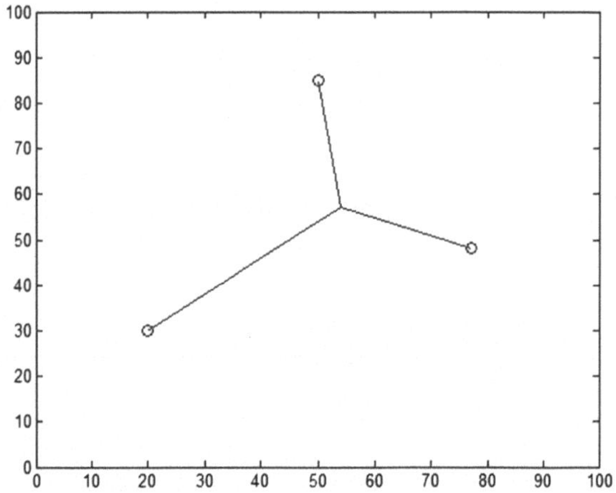

Fig. 3 Steiner solution for three random points

Fig. 4 Notations

$N1_{org}$, $N2_{org}$ – Set of original nodes of network 1 and 2 respectively

L_{N1org}, L_{N2org} – Number of elements in $N1_{org}$ and $N2_{org}$ respectively

$N1_{edge}$, $N2_{edge}$ – Set of edge nodes of network 1 and 2 respectively

L_{N1edge}, L_{N2edge} – Number of elements in $N1_{edge}$ and $N2_{edge}$ respectively

T – Target set containing the three nodes which will give final solution

B_{final} – Brigde node

$R_{edge}(x)$ – distance between node x and line(x=50)

R(x,y) – distance between node x and node y

$^{n}C_{2}$ – all possible selections of 2 nodes from set whose length is denoted by n

steinerlocation(x,y,z) – returns the location of steiner point solution for nodes x,y,z

```
FOR  i = 1 to L_N1org          // step I
   FOR j = 1 to L_N1org
      IF  x-coordinate of N1_org(j) > x-coordinate of
N1_org(i)
          IF  R(N1_org(j),  N1_org(i)) < R_edge(N1_org(i))
              do not make  N1_org(i) element of N1_edge
              flag =1
              BREAK
          END IF
      END IF
      IF flag == 0
         make  N1_org(i) element of N1_edge
      ELSE     node i an edge node
      END IF
   END FOR
END FOR
REPEAT  the above process for network 2
FOR i = 1 to L_N1edge                    //  steps II and III
    IF  minsum > steinercost(N1_edge(i), N2_edge( ^{LN2edge}C_2 ))
        minsum = steinercost(N1_edge(i), N2_edge( ^{LN2edge}C_2 ))
        REPLACE  elements of set T with N1_edge(i), N2_edge(
^{LN2edge}C_2 )
    END IF
END FOR
FOR i = 1 to L_N2edge
   IF  minsum > steinercost(N2_edge(i), N1_edge( ^{LN2edge}C_2 ))
       minsum = steinercost(N2_edge(i), N1_edge( ^{LN2edge}C_2 ))
       REPLACE  elements of set T with N2_edge(i), N1_edge(
^{LN2edge}C_2 )
   END IF
END FOR
Bfinal  = steinerlocation(T)                  // step IV


Finding the cost of a particular set of three points -
Steinercost(x,y,z)
P =  steinerlocation(T)
cost = distance(x,P) + distance(y,P) + distance(x,P)
RETURN cost
```

Fig. 5 High-level pseudocode of the proposed algorithm

4 Results and Discussion

Important points to be noted:

- The motivation to find edge nodes is based on the understanding that any communication with the other network will have to take place through the edge node. Otherwise, the path of communication will not be shortest.
- The Steiner tree algorithm for three points gives the location of Steiner point which can be used to interconnect all the three points with the shortest possible paths.
- For the sake of simplicity, the common region is assumed to be a single line ($x = 50$). In reality, the common region may be a wider area.

The following solution was obtained for the placement of nodes as shown in Fig. 6.

This (Fig. 8) is the required solution for optimum placement of bridge node. It is obtained by applying the Steiner algorithm among all combinations of the edge nodes taken three at a time (with at least one from each network). For the purpose of better visualization of the network, a minimum spanning tree may be plotted for any node as root node. This gives shortest path communication from that node to any other node (Figs. 7 and 9).

Assuming all nodes generate traffic equally, we may find the shortest path between any two nodes for all nodes and subsequently find the sum. Adding the above distance values gives a total of **1656.6**. Now we shall compare this result with the alternate approaches that were mentioned earlier in the paper.

Fig. 6 Random placement of 8 nodes

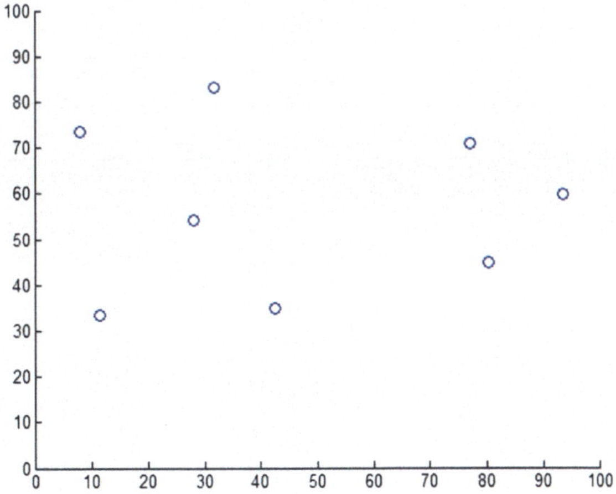

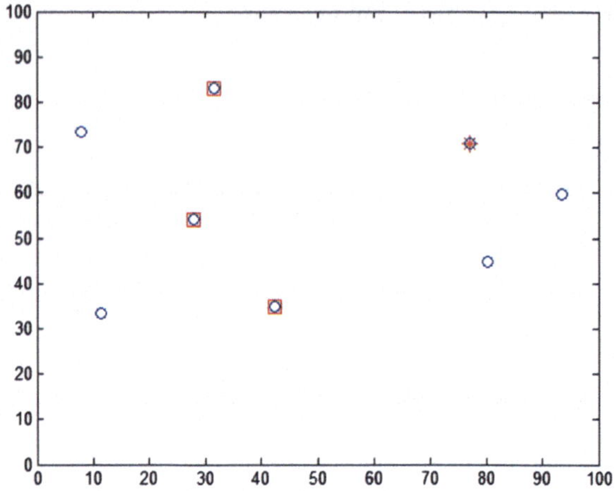

Fig. 7 Determination of edge nodes shown in red square for network 1 and star for network 2

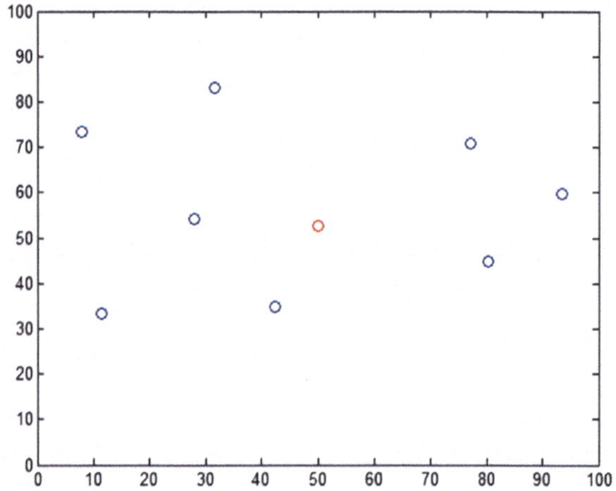

Fig. 8 Location of bridge node indicated by red circle along $x = 50$ line

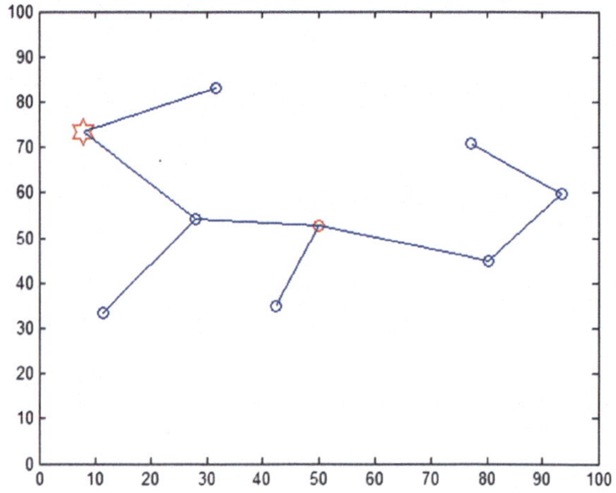

Fig. 9 Minimum spanning tree with root node indicated by star

5 Comparative Analysis

The method used in this paper presents two unique approaches:

1. Using edge nodes

As mentioned previously, the idea of using edge nodes is based on the understanding that any communication with the other network will have to occur through these nodes and hence all optimization efforts with respect to the placement of bridge nodes should be focussed only on these nodes. Otherwise, the Steiner algorithm may yield a different solution which may not be the shortest possible path. The following example illustrates this point.

The same set of nodes has been used as above. However, no edge nodes were identified. Instead, the Steiner algorithm was applied to all possible combinations to find the set of three points which yield the minimum sum of paths (brute force approach).

This approach yields a different solution as compared to Fig. 6. This is because, in this particular case, the Steiner point is obtained as a result of applying it on the two nodes of network 1 having lowest y value and 1 node of network 2 having lowest y value. While these 3 points may yield the lowest sum of paths compared to any other set of 3 points, they do not give the best result overall.

Deviations from Fig. 10 can be observed in the nodes with source = 9 (Figs. 11, 12, 13, 14, 15, 16, and 17).

Adding the above distances gives a total of 1685.8 which is 29.2 more than the previous solution using edge nodes. Thus, the proposed solution tends to give better results than the brute force approach.

2. Applying Steiner algorithm for 3 nodes at a time

The Steiner Tree algorithm can be applied for any number of points. However, finding the solution for any number of points is very difficult. Assuming the solution is available from any of the available published algorithms [13], the results are still not the best. This is because the problem at hand is not a pure Steiner Tree problem.

- There is a constraint on the placement of the bridge node. It must be at $x = 50$.
- We must not change the structure of the existing networks by adding new nodes, etc.

The following example illustrates this point. The Steiner Tree for given 8 points was found using one of the available published algorithms. The point where the tree intersects the line $x = 50$ was chosen as location of bridge node.

Deviations from Fig. 10 can be observed in the nodes with source = 9.

Fig. 10 Shortest path distance for all pairs of nodes. The nodes are indexed in the order in which the input was fed. The last node (9) is the Steiner node

Source	Destination	Distance
2	1	25.62
3	1	27.82
3	2	29.18
4	1	40.20
4	2	53.68
4	3	26.57
5	1	51.80
5	2	49.44
5	3	24.14
5	4	31.14
6	1	69.17
6	2	47.02
6	3	51.83
6	4	75.58
6	5	49.88
7	1	86.58
7	2	66.03
7	3	65.67
7	4	86.14
7	5	56.66
7	6	19.77
8	1	77.81
8	2	61.89
8	3	53.13
8	4	69.83
8	5	39.07
8	6	26.22
8	7	19.87
9	1	46.83
9	2	35.37
9	3	22.04
9	4	43.21
9	5	19.47
9	6	32.50
9	7	43.97
9	8	31.34

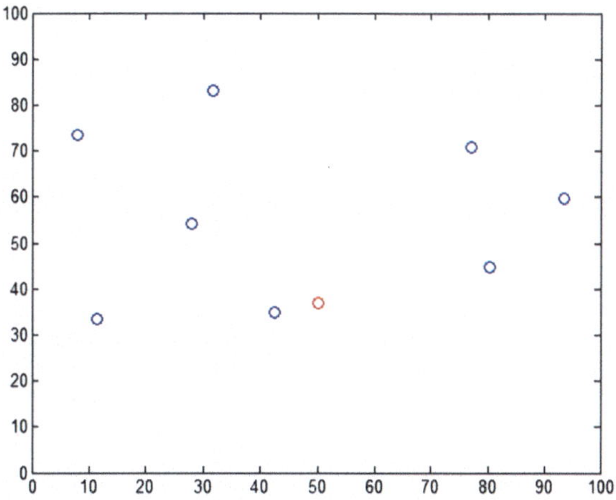

Fig. 11 Location of bridge node indicated by red circle along $x = 50$ line

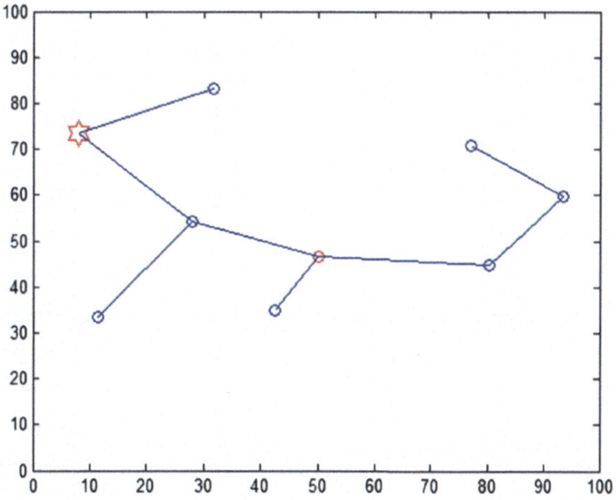

Fig. 12 Minimum spanning tree with same root node indicated by star

Adding the above values gives a total of 1662.9 which is 6.3 more than the original solution that was first presented in this paper. Thus, the proposed solution tends to give better results than using the pure Steiner Tree solution approach.

The three approaches—proposed solution (PS), brute force (BF) and pure Steiner tree (PST)—have been tested on various input arrangements. Five such arrangements have been shown in Fig. 18. For each input, the sum of shortest path distances for all pairs of nodes was calculated using each of the three approaches. The distances are then scaled to the value of the proposed solution (PS) so that they can be easily compared.

Thus, for most cases, the proposed solution yields a better solution; however, there may be certain arrangements for which the three approaches may give same result as in case of arrangement (e) (Figs. 19 and 20).

Fig. 13 Shortest path distance for all pairs of nodes. Row indicates source and column indicates destination. The diagonal and upper right corners of the matrix are to be ignored

Source	Destination	Distance
2	1	25.62
3	1	27.82
3	2	29.18
4	1	40.20
4	2	53.68
4	3	26.57
5	1	51.80
5	2	49.44
5	3	24.14
5	4	31.14
6	1	69.17
6	2	47.02
6	3	51.83
6	4	75.58
6	5	49.88
7	1	86.58
7	2	66.03
7	3	65.67
7	4	86.14
7	4	86.14
7	5	56.66
7	6	19.77
8	1	77.81
8	2	61.89
8	3	53.13
8	4	69.83
8	5	39.07
8	6	26.22
8	7	19.87
9	1	**55.76**
9	2	**49.76**
9	3	**28.00**
9	4	**38.74**
9	5	**7.74**
9	6	**43.45**
9	7	**49.09**
9	8	**31.33**

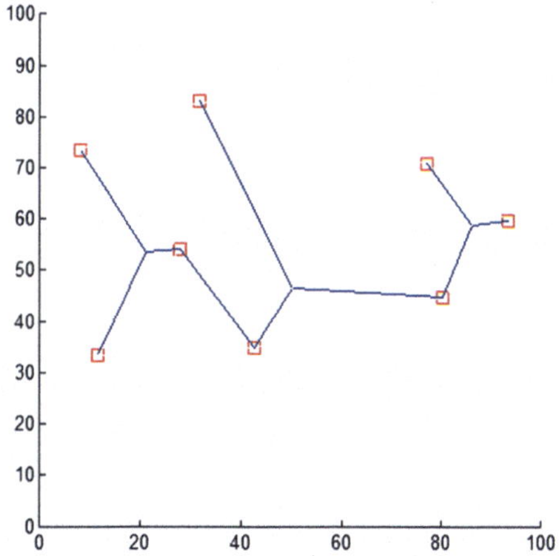

Fig. 14 Minimum Steiner tree for given 8 points. The tree intersects at $x = 50$ at $y = 46.5$

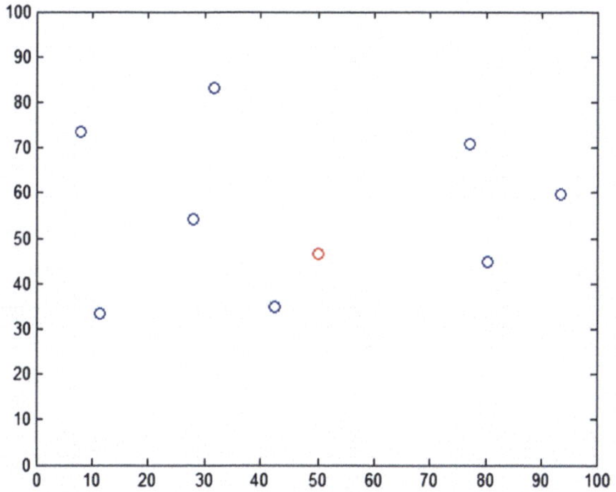

Fig. 15 Location of bridge node indicated by red circle along $x = 50$ line. Compare with Figs. 6 and 9

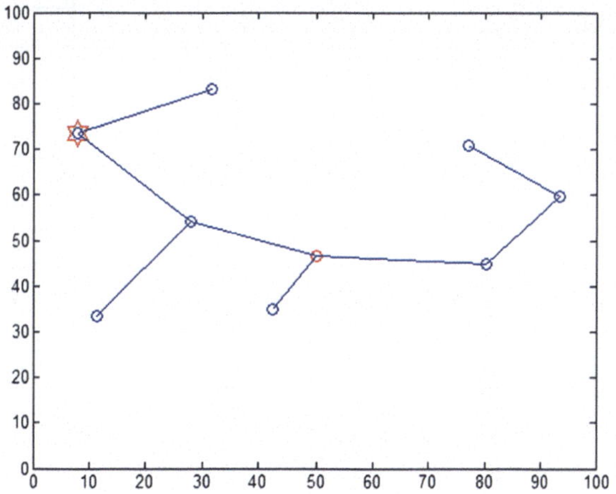

Fig. 16 Minimum spanning tree with same root node indicated by star

Fig. 17 Shortest path distance for all pairs of nodes. Row indicates source and column indicates destination. The diagonal and upper right corners of the matrix are to be ignored

Source	**Destination**	**Distance**
2	1	25.62
3	1	27.82
3	2	29.18
4	1	40.20
4	2	53.68
4	3	26.57
5	1	51.80
5	2	49.44
5	3	24.14
5	4	31.14
6	1	69.17
6	2	47.02
6	3	51.83
6	4	75.58
6	5	49.88
7	1	86.58
7	2	66.03
7	3	65.67
7	4	86.14
7	5	56.66
7	6	19.77
8	1	77.81
8	2	61.89
8	3	53.13
8	4	69.83
8	5	39.07
8	6	26.22
8	7	19.87
9	1	**49.99**
9	2	**41.00**
9	3	**23.32**
9	4	**40.73**
9	5	**13.77**
9	6	**36.45**
9	7	**45.42**
9	8	**30.34**
9	8	**31.33**

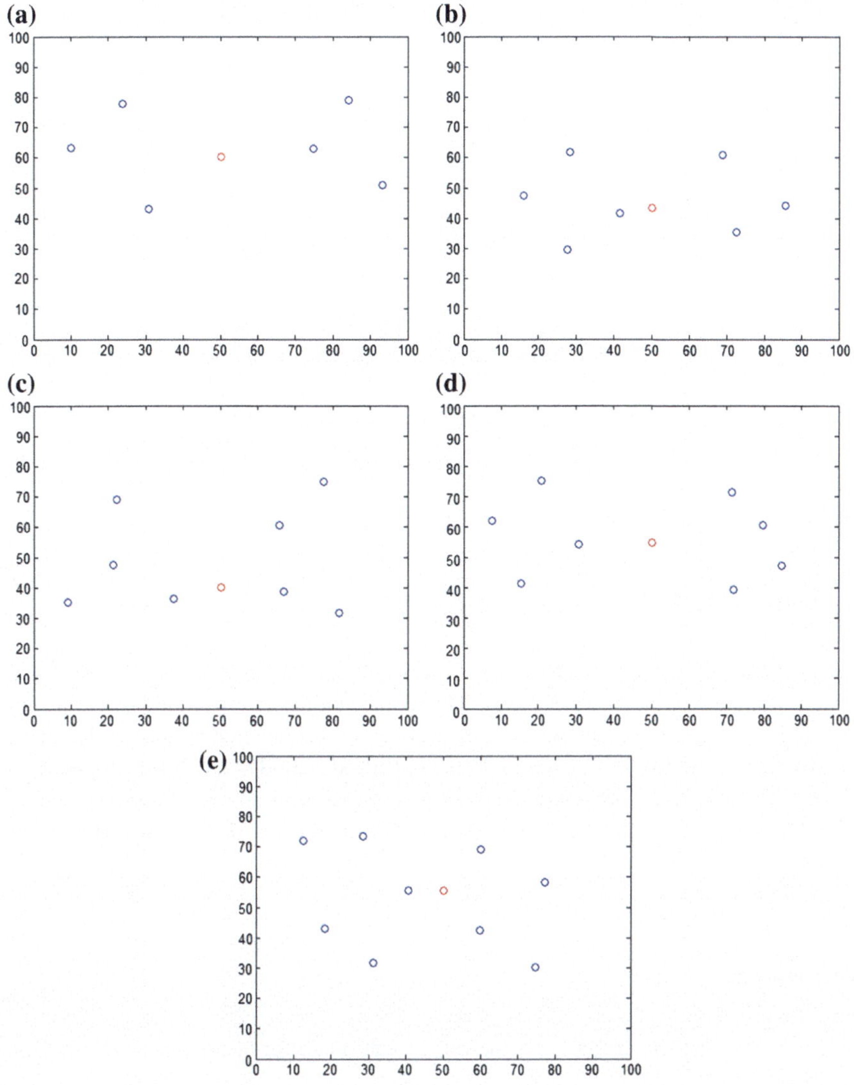

Fig. 18 Input arrangements. **a** 6 nodes, **b** 7 nodes, **c** 8 nodes, **d** 8 nodes, **e** 9 nodes

Arrangement	PS	BF	PST	PS (scaled)	BF (scaled)	PST (scaled)
a	948.37	956.6	950.13	100	100.86	100.18
b	987.21	990.22	990.21	100	100.3	100.3
c	1455.7	1461.6	1456.1	100	100.45	100.02
d	1498.5	1500.5	1502.2	100	100.13	100.24
e	1630	1630	1630	100	100	100

Fig. 19 Shortest path distances

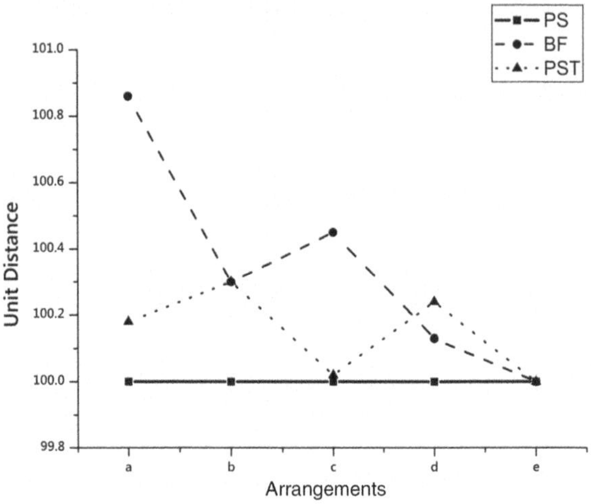

Fig. 20 Comparison of approaches

6 Conclusion

The challenge of interconnecting two or more independent networks has assumed greater significance in modern times. This paper addresses the challenge in a scenario where there is constraint on the number of bridge nodes and its location. A combination of MST solution and concept of edge nodes has been proposed in this paper. The results obtained were also compared with alternate approaches like brute force approach where it was observed that the proposed solution yields better results.

References

1. Derek RD, Michael LO (2002) Two heuristics for the Euclidean Steiner tree problem. J Glob Optim 95–106
2. Gröpl C, Hougardy S, Nierhoff T, Prömel H (2001) Approximation algorithms for the Steiner tree problem in graphs. In: Cheng X, Du D-Z (eds) Steiner trees in industry. Kluwer, Norwell, pp 235–279
3. Zelikovsky A (1993) An 11/6-approximation algorithm for the network Steiner problem. Algorithmica 9:463–470
4. Berman P, Ramaiyer V (1994) Improved approximations for the Steiner tree problem. J Algorithms 17:381–408
5. Karpinski M, Zelikovsky A (1997) New approximation algorithms for the Steiner tree problems. J. Comb. Optim. 1:47–65
6. Liu L, Song Y, Zhang H, Ma H, Vasilakos AV (2015) Physarum optimization: a biology-inspired algorithm for the Steiner tree problem in networks. IEEE Trans Comput 64(3):818–831. https://doi.org/10.1109/tc.2013.229

7. Cong J, Kahng A, Leung K (1998) Efficient algorithms for the minimum shortest path Steiner arborescence problem with applications to VLSI physical design. IEEE Trans Comput Aided Des Integr Circ Syst 17(1):24–39
8. Cheng X, Du D (2001) Steiner trees in industries. Springer, Berlin
9. Oliveira C, Pardalos P (2005) A survey of combinatorial optimization problems in multicast routing. Comput Oper Res 32(8):1953–1981
10. Wang W, Li X, Wang Y (2004) Truthful multicast in selfish wireless networks. In: Proceedings of the 10th Annual International Conference on Mobile Computing and Networking (Mobi-Com'04), Sept 26/Oct 1, 2004, pp 245–259
11. Misra S, Hong S, Xue G, Tang J (2010) Constrained relay node placement in wireless sensor networks: formulation and approximations. IEEE/ACM Trans Netw 18(2):434–447
12. Mao X, Miao X, He Y, Zhu T, Wang J, Dong W, Li X, Liu Y (2012) CitySee: urban CO_2 monitoring with sensors. In: Proceedings of IEEE International Conference on Computer Communications (INFOCOM), pp 1611–1619
13. do Nascimento MZ, Batista VR, Coimbra WR (2015) An interactive programme for weighted Steiner trees. J Phys (Conference Series 574 (2015)) 012073. https://doi.org/10.1088/1742-6596/57

Energy-Efficient Elliptic Curve Cryptography-Based DTLS Key Establishment Protocol for IoT Communication

P. N. V. Karthik, R. Rajashree, Vijayakumar Perumal and Ganesan Veerappan

Abstract Security is the most important aspect in the Internet of Things (IoT) network. Datagram transport layer security (DTLS) protocol helps in achieving secure communication in IoT network. To run the DTLS protocol, key establishment should take place between communicating devices. Key establishment can be done by using either symmetric keys or asymmetric keys. Using symmetric key (preshared key) mode perfect secrecy and security cannot be achieved. If symmetric key is compromised, then there may be a chance of decrypting the messages of the previous sessions by the attacker. In order to overcome above-mentioned drawbacks and to increase the level of security an efficient elliptic curve cryptography (ECC)-based DTLS key establishment protocol for IoT communication is proposed. In the proposed method, ECC plays a major role in providing security. The proposed key establishment protocol is implemented in MATLAB, and the performance of proposed protocol with traditional key establishment protocol in terms of different simulation parameters is compared.

Keywords Internet of things (IoT) · Elliptic curve cryptography (ECC)
Security · Key establishment · Datagram transport layer security (DTLS)

P. N. V. Karthik · V. Perumal (✉)
VIT University, Chennai, India
e-mail: vijayrgcet@gmail.com; vijaya.kumar@vit.ac.in

P. N. V. Karthik
e-mail: karthik888.pnv@gmail.com

R. Rajashree · G. Veerappan
Sathyabama University, Chennai, India
e-mail: Rajashree.ece@gmail.com

G. Veerappan
e-mail: vganesh1711@gmail.com

A. M. Zungeru et al. (eds.), *Wireless Communication Networks and Internet of Things*, Lecture Notes in Electrical Engineering 493,
https://doi.org/10.1007/978-981-10-8663-2_7

1 Introduction

Internet of Things is a network where different objects are interconnected and able to communicate, exchange data with each other. The data which is to be exchanged must be protected by using some cryptographic techniques. Initially to start session between client and server, key establishment should take place between them. As the client has only temporary relationship with the server giving the server's secret key in symmetric key mode and public key mode is not suitable. To solve this problem, new protocols have been developed.

Newly developed protocols use a trusted third party technique (TTP). TTP instead of giving secret key of the server to the client, it gives temporary keys. The proposed protocol uses trusted third party (TTP) technique, and it consists of two approaches to establish trust relationship between client and resource server. In first approach, the trusted third party establishes a new public key without disclosing the resource server's secret key. In the second approach, the trusted third party issues a certificate to the client as well as resource server. The client and resource server use this certificate during key establishment process and generate session keys. Further sections of this paper are categorized as follows. Second section is about the existing protocol. Third section gives details about the proposed protocol. In the fourth section, performance analysis is presented. Security analysis is mentioned in section five. Conclusion of the proposed protocol is presented in sixth section. In seventh section, references are mentioned [1].

2 Existing Key Establishment Protocol

AES is used in existing key establishment protocol, and it consists of two approaches to establish a relation between client and server. In first approach, symmetric keys are used to establish a secure DTLS connection with resource server but symmetric keys have few drawbacks. In symmetric key mode, the keys have to be shared to all the parties involved before starting handshake. Therefore, perfect forward secrecy cannot be achieved and there may be a chance of decrypting the messages of previous sessions by an attacker. If symmetric keys are used, then any one holding a key can act like other person because same key is shared to all of them. Second approach is to use raw public keys. Instead of certificates, raw public keys can be used to establish a relation with a resource server. But some modifications have to be made to raw public keys to use them for secure DTLS connections [2].

3 Proposed ECC-Based DTLS Key Establishment Protocol

Elliptic curve cryptography (ECC) is used in our protocol to increase the level of security. Using ECC, high level of security can be obtained by using smaller key lengths. ECC helps in achieving faster processing speed and reduces the communication complexity and key storage requirements. The proposed protocol consists of two approaches to establish trust relationship between client and resource server. In first approach, the trusted third party establishes a new public key without disclosing the resource server's secret key. In the second approach, the trusted third party issues a certificate to the client as well as resource server. The client and resource server use this certificate during key establishment process, generate session keys, and verify them [3].

3.1 Derivation of Public Keys

First an elliptic curve is defined and a base point 'p' is selected from the set of generated elliptic points. As shown in Fig. 1, client selects a random number 'g_C'. Here, 'g_C' is client's temporary private key. Client then finds temporary public key 'G_C', it is formulated as $G_C = g_C \times p$. Trust anchor selects a random number 'g_{TA}'. Here, 'g_{TA}' is temporary private key of trust anchor. Trust anchor then computes its temporary public key, and it is given as $G_{TA} = g_{TA} \times p$. Similarly, resource server selects a random number 'g_{RS}' and computes temporary public key given by $G_{RS} = g_{RS} \times p$. Client then sends its temporary public key 'G_C' and identity of client 'ID_C' to trust anchor. Similarly, resource server sends its 'G_{RS}' and identity of resource server 'ID_{RS}' to trust anchor. After receiving the temporary public keys, identities of both client and resource server the trust anchor will verify client's identity and resource server's identity [4].

Trust anchor selects a temporary key pair (q_{TA}, Q_{TA}) and computes the elliptic points of client $B_C = G_C + G_{TA}$ and resource server $B_{RS} = G_{RS} + G_{TA}$. Trust anchor then applies hash function $H(Q_{TA}, ID_C)$ to derive integer 'e_C' and $H(Q_{TA}, ID_{RS})$ to derive integer 'e_{RS}'. Trust anchor then computes client's temporary private-key data $S_C = g_{TA} \times e_C + q_{TA} \pmod{n}$ and RS temporary private-key data $S_{RS} = g_{TA} \times e_{RS} + q_{TA} \pmod{n}$. Trust anchor computes client's public key $Q_C = e_C \times B_C + Q_{TA}$ and resource server's public key $Q_{RS} = e_{RS} \times B_{RS} + Q_{TA}$ and sends S_C, B_C to client and S_{RS}, B_{RS} to resource server. After receiving S_C, B_C from TA, client computes $H(ID_C, Q_{TA})$ and derives an integer 'e_C'. After receiving S_{RS}, B_{RS} from TA, resource server computes $H(ID_{RS}, Q_{TA})$ and derives an integer 'e_{RS}'. In the next step, client computes its private key 'q_C'. Private key is formulated as $q_C = S_C + g_C \times e_C \pmod{n}$ and its public key $Q_C = q_C \times p$. Then resource server computes its private key $q_{RS} = S_{RS} + g_{RS} \times e_{RS} \pmod{n}$ and its public key $Q_{RS} = q_{RS} \times p$. In the next step, client and resource server find their

Client (C)	Trust anchor (TA)	Resource server (RS)
Choose random value g_C	Choose random value g_{TA}	Choose random value g_{RS}
Compute $G_C = g_C \times p$	**Compute** $G_{TA} = g_{TA} \times p$	**Compute** $G_{RS} = g_{RS} \times p$
Send G_C, ID_C to TA →	**Compute** $B_C = G_C + G_{TA}$	← **Send** G_{RS}, ID_{RS} to TA
	Compute $B_{RS} = G_{RS} + G_{TA}$	
	Compute $H(Q_{TA}, ID_C) = e_C$	
	Compute $H(Q_{TA}, ID_{RS}) = e_{RS}$	
	Compute $S_C = g_{TA} \times e_C + q_{TA}$ (mod n)	
	Compute $S_{RS} = g_{TA} \times e_{RS} + q_{TA}$ (mod n)	
	Compute $Q_C = e_C \times B_C + Q_{TA}$	
	Compute $Q_{RS} = e_{RS} \times B_{RS} + Q_{TA}$	
Compute $e_C = H(Q_{TA}, ID_c)$	← **Send** S_C, B_C to client; **Send** S_{RS}, B_{RS} to RS →	**Compute** $e_{RS} = H(Q_{TA}, ID_{RS})$
Compute $q_C = S_C + g_C \times e_C$ (mod n)		**Compute** $q_{RS} = S_{RS} + g_{RS} \times e_{RS}$ (mod n)
Compute $Q_C = q_C \times p$		**Compute** $Q_{RS} = q_{RS} \times p$
Compute $Q_C^1 = e_C \times B_C + Q_{TA}$		**Compute** $Q_{RS}^1 = e_{RS} \times B_{RS} + Q_{TA}$
Verify $Q_C = Q_C^1$		**Verify** $Q_{RS} = Q_{RS}^1$

Fig. 1 Sequence of steps required for the derivation of public keys

respective reconstructed public keys Q_C^1, Q_{RS}^1 given by $Q_C^1 = e_C \times B_C + Q_{TA}$ and $Q_{RS}^1 = e_{RS} \times B_{RS} + Q_{TA}$. If $Q_C = Q_C^1$, then the public key is given as (q_C, Q_C) otherwise it rejects the key pair. Similarly if $Q_{RS} = Q_{RS}^1$, the key pair is given as (q_{RS}, Q_{RS}) otherwise it rejects the key pair [5].

3.2 Certificate Generation and Derivation of Session Keys

In the second approach as shown in Fig. 2, the trust anchor selects a random number 'g_{TA}' as its private key. Trust anchor then computes its public key $G_{TA} = g_{TA} \times p$. Client selects a random number 'g_C' and computes temporary public key $G_C = g_C \times p$. Client then sends 'G_C', identity of client 'ID_C' to trust anchor. Trust anchor verifies the identity of client. Then trust anchor selects a temporary key pair (q_{TA}, Q_{TA}) and then finds the elliptic point $B_C = G_C + G_{TA}$. The certificate 'I_C' is generated by trust anchor and it consists of trust anchor's public key 'Q_{TA}', identity of client 'ID_C', elliptic point 'B_C' certificate expiry date 't_C', random nonce 'N_C', time stamp 'T_{SC}'. Trust anchor then applies hash function $H(I_C)$ to derive an integer 'e_C'. In the next step, trust anchor computes private-key data $S_C = g_{TA} \times e_C + q_{TA}$ (mod n) and client's public key $Q_C = e_C \times B_C + Q_{TA}$. Trust anchor sends I_C, B_C back to client. Client then reconstructs the public key $Q_C^1 = e_C \times B_C + Q_{TA}$. If $Q_C = Q_C^1$, client accepts the certificate and the key pair is taken as (q_C, Q_C). If $Q_C \neq Q_C^1$, then client will reject the certificate. By using the same process as mentioned above, resource server will obtain a certificate from trust anchor and the key pair will be taken as (q_{RS}, Q_{RS}) [6].

Client (C)	Trust anchor (TA)
Select a random number g_C	Select random number g_{TA}
Compute $G_C = g_C \times p$	**Compute** $G_{TA} = g_{TA} \times p$
Send G_C, ID_C →	Generate random nonce N_C,
	Certificate expiry t_C, Time stamp T_{SC}
	Compute $B_C = G_C + G_{TA}$
	Compute $I_C = (ID_C, t_C, N_C, T_{SC}, B_C, Q_{TA})$
	Compute $H(I_C) = e_C$
	Compute $S_C = g_{TA} \times e_C + q_{TA} \pmod n$
Compute $H(I_C) = e_C$	**Compute** $Q_C = e_C \times B_C + Q_{TA}$
Compute $q_C = S_C + g_C \times e_C \pmod n$	← **Send** S_C, I_C
Compute $Q_C = q_C \times p$	
Compute $Q_C^1 = e_C \times B_C + Q_{TA}$	
Compute $Q_C = Q_C^1$?	

Fig. 2 Implicit certificate generation process of client

As shown in Fig. 3, client and resource server share a secret key 'Q_K'. Client chooses a random number 'g_C' and computes $G_C = g_C \times p$. Where 'g_C' is client's temporary private key and 'G_C' is client's public key. Resource server chooses a random number 'g_{RS}' and computes $G_{RS} = g_{RS} \times p$. Where 'g_{RS}' is resource server's temporary private key and 'G_{RS}' is resource server's public key. Client computes g_C (G_{RS}) and it is assigned to c. Client then computes $f = h(ID_C, Q_K)$, $e = E_f(G_C)$ and sends f, e, certificate I_C to resource server. Resource server extracts ID_C, t_C, N_C, B_C, T_{SC}, Q_{TA} form the certificate I_C and checks the validity of certificate expiry 't_C'. RS sends certificate 'I_S' to client. After receiving, client verifies the certificate expiry date of resource server. RS then computes $f = h(ID_C, Q_K)$, $G_C = D_f(e)$, and $c = g_{RS}(G_C)$. Resource server generates the session key

Client (C)	Resource server (RS)
Choose random number 'g_C'	Choose random number 'g_{RS}'
Compute $G_C = g_C \times p$	**Compute** $G_{RS} = g_{RS} \times p$
Shared secret key is 'Q_K'	'Q_K' is Shared secret key
Compute $c = g_C(G_{RS})$, $f = h(ID_C, Q_K)$,	$I_S = (ID_{RS}, t_S, N_C, T_{RS}, B_{RS}, Q_{TA})$
$e = E_f(G_C)$	← **Send** I_S
$I_C = (ID_C, t_C, N_C, T_{SC}, B_C, Q_{TA})$	
Send I_C →	Extracts ID_C, N_C, T_{SC} and checks the validity of 't_C'
Checks the validity of 'I_S'	**Compute** $f = h(ID_C, Q_K)$, $G_C = D_f(e)$
	Compute $c = g_{RS}(G_C)$
	Compute $SK_{RS} = h(c, G_C, N_C, ID_C)$
	Compute $M_{RS} = h(SK_{RS}, Q_K, G_{RS})$
Compute $SK_C = h(c, G_C, N_C, ID_C)$	← **Send** M_{RS}
Compute $M_{RS} = h(SK_C, Q_K, G_{RS})$	
Compute $M_C = h(SK_C, Q_K, G_{RS})$	
Send M_C →	**Verify** $M_C = h(SK_{RS}, Q_K, G_{RS})$

Fig. 3 Sequence of steps required for the session key generation of client and resource server

Table 1 Comparison table for DTLS key establishment protocol using AES and ECC

S. No.	Phases	Existing system—AES		Proposed system—ECC		Reduced processing time (%) for ECC
		Key size (bits)	Processing time (s)	Key size (bits)	Processing time (s)	
1	Public key derivation	40	0.062875	40	0.039156	37
2	Client certificate generation	40	0.059363	40	0.047492	20
3	Resource server certificate generation	40	0.067187	40	0.043596	35
4	Session key generation	40	0.051358	40	0.029429	42

$SK_{RS} = h(c, G_C, N_C, ID_C)$ and authenticator $M_{RS} = h(SK_{RS}, Q_K, G_{RS})$. Resource server then sends 'M_{RS}' to client. Client then computes session key $SK_C = h(c, G_C, N_C, ID_C)$ and checks whether $M_{RS} = h(SK_C, Q_K, G_{RS})$. If yes, client believes that resource server is authenticated and uses the key 'SK_C' to communicate with resource server. Then client computes the authenticator $M_C = h(SK_C, Q_K, G_{RS})$ and sends 'M_C' to resource server. After receiving M_C, resource server checks whether $M_C = h(SK_{RS}, Q_K, G_{RS})$. If yes, resource server believes that client is authenticated and uses session key 'SK_{RS}' to communicate with the client securely [7, 8].

4 Performance Analysis

Proposed protocol is implemented in MATLAB and the performance of both the approaches is analyzed in this section. Parameters such as key size and processing time are taken into consideration. Reduced processing time during all the phases for the proposed system when compared with existing AES system is mentioned in Table 1. Respective figures are plotted to compare the total processing time taken during derivation of keys by using ECC and AES (Figs. 4, 5, 6 and 7) [9].

5 Security Analysis

ECC used in the proposed protocol helps to resist against various attacks such as replay attack, modification attack, forgery attack, man-in-the-middle attack, node compromise attacks and also provides identity privacy, perfect forward and backward secrecy for both the approaches [10].

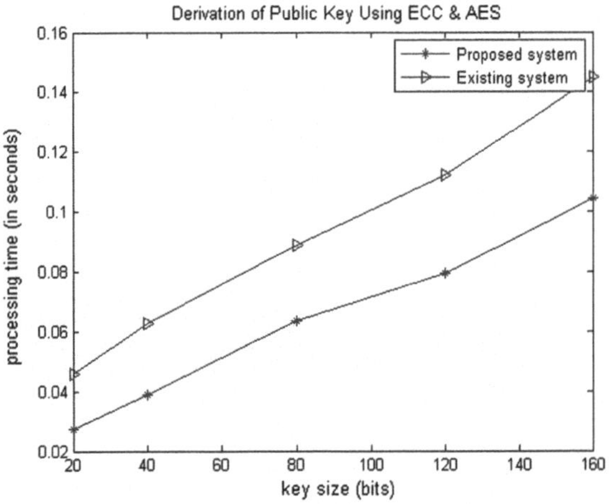

Fig. 4 Comparison of total processing time taken for the derivation of public key using ECC and AES

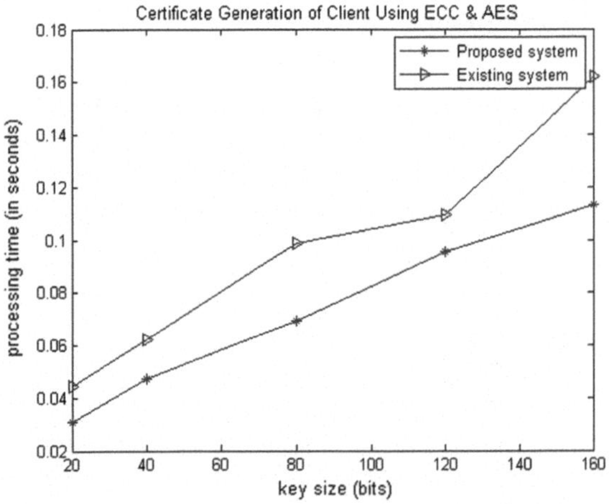

Fig. 5 Comparison of total processing time taken for the certificate generation of client using ECC and AES

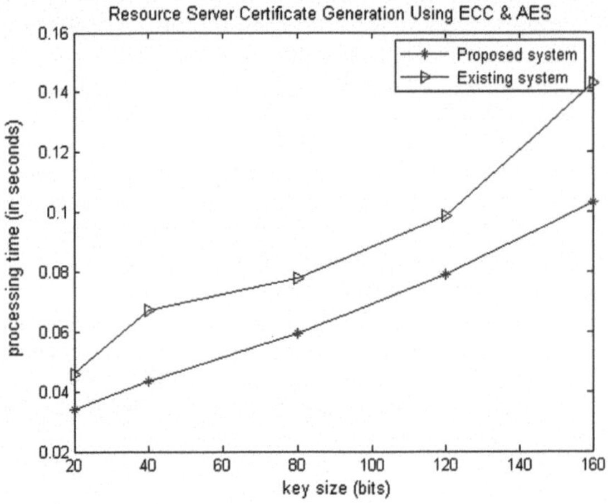

Fig. 6 Comparison of total processing time taken for the certificate generation of resource server using ECC and AES

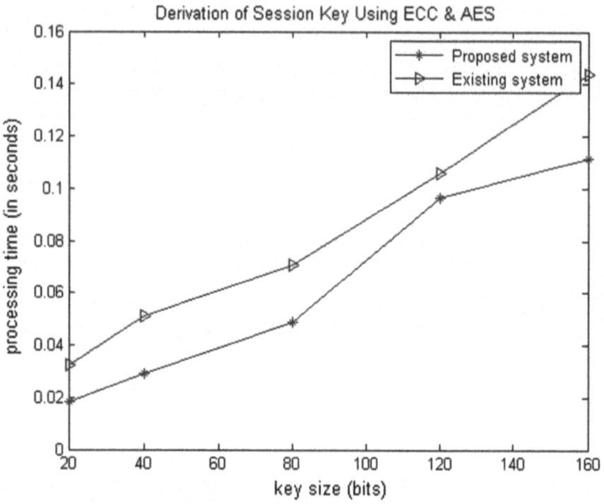

Fig. 7 Comparison of total processing time taken for the session key generation using ECC and AES

6 Conclusion

The proposed key establishment protocol is highly secure because of the security provided by ECC, and it can resist against various attacks. ECC creates faster, smaller, and efficient keys in a considerably less amount of time when compared with traditional key establishment protocols.

References

1. Tiloca M, Gehrmann C, Seitz L (2016) On improving resistance to denial of service and key provisioning scalability of the DTLS handshake. Int J Inf Secur
2. Borgia E (2014) The Internet of things vision: key features, applications and open issues. Comput Commun J
3. Shen H, Shen J, Khan MK, Lee J-H (2016) Efficient RFID authentication using elliptic curve cryptography for the Internet of things. J Wireless Peers Commun
4. Zhang Y, Shen Y, Wang H, Yong J, Jiang X (2016) On secure wireless communications for IoT under eavesdropper collusion. IEEE Trans Autom Sci Eng 13(3)
5. Aldosari W, El Taeib T (2015) Secure key establishment for device-to-device communications among mobile devices. Int J Eng Res Rev 3(2)
6. Sheng Z, Wang H, Yin C, Hu X, Yang S, Leung VCM (2015) Lightweight management of resource constrained sensor devices in Internet of things. IEEE Internet Things J 2(5)
7. Kothmayr T, Schmitt C, Hu W, Brunig M, Carle G (2013) DTLS based security and two-way authentication for the Internet of things. Ad Hoc Netw
8. Chavan AA, Nighot MK (2014) Secure CoAP using enhanced DTLS for Internet of things. Int J Innov Res Comput Commun Eng 2(12)
9. Premnath SN, Haas ZJ (2015) Security and privacy in the Internet of things under time and budget limited adversary model. IEEE Wireless Commun Lett 4(3)
10. Raza S, Seitz L, Sitenkov D, Selander G (2016) S3K: scalable security with symmetric keys—DTLS key establishment for the Internet of things. IEEE Trans Autom Sci Eng 13(3)

Monitoring Sensor Nodes with COOJA Simulator

U. N. V. P. Rajendranath and V. Berlin Hency

Abstract Usually, the sensors are connected to the gateway units in the Internet of Things sensory environment and this gives birth to many sensors based real-time applications. Sensors are deployed in a remote location to monitor the temperature and light intensity in industries and these sensors are connected to the border router by CoAP protocol. The information of sensors is displayed in the Webpage periodically for monitoring the parameters. The border router is connected to the Internet using IPv6 Internet protocol. The simulation can be done by using COOJA simulator and the parameters obtained are continuously monitored by analyzing the power consumption of nodes.

1 Introduction

Nowadays, research in wireless sensor networks focused mainly on the routing protocol, MAC protocol, and the sensor nodes location management. By employing gateway unit in between the wireless sensor nodes and cloud, it provides data storage and aggregation and also the protocol conversion.

As know that wireless sensor node consists of a microcontroller unit, a transceiver, memory, timer, and analog to digital converter. Wireless sensor network consists of several sensing nodes deployed in constrained environment. The sensing devices individually form a network for transporting the data. The battery-operated devices or a solar panel can supply only the limited amount of power to the sensing devices. For finding the suitable operating system for sensor nodes, the programmer must find the lightweight mechanisms that provide ample enough environment execution while considering the limitations of constrained environments. For achieving the high quality and fault tolerance and also considering the QoS

U. N. V. P. Rajendranath (✉) · V. Berlin Hency
School of Electronics Engineering Department, VIT University, Chennai, India
e-mail: udathunv.pandurangarajendranath2015@vit.ac.in

V. Berlin Hency
e-mail: berlinhency.victor@vit.ac.in

A. M. Zungeru et al. (eds.), *Wireless Communication Networks and Internet of Things*, Lecture Notes in Electrical Engineering 493,
https://doi.org/10.1007/978-981-10-8663-2_8

parameters in wireless sensor networks, a multi-hop communication can be employed in the dense deployment of sensor nodes.

The operating system CONTIKI specially designed for environments under several constraints. CONTIKI [1] supports various microcontroller architectures; among them, MSP430 and Atmel AVR use C language to port.

The operating system CONTIKI is based on the modular architecture. The lightweight event scheduler of CONTIKI sends events to the processes which are in running state. The execution of these processes is triggered by the events forwarded by the kernel to the processes. To avoid race conditions, polling must be used.

Two types of events are supported by the CONTIKI OS: one is a synchronous event and another one is an asynchronous event. In synchronous event, events are dispatched immediately for completing the task scheduled. On another hand, asynchronous events in which tasks are queued for some time and dispatched for task processing. The high-priority events are scheduled by preempting the asynchronous event for completing the high-priority event to execute.

COOJA [2] is a CONTIKI network emulator. COOJA simulates the large and small networks of CONTIKI motes. The CONTIKI motes are z motes, sky motes, ESB motes, Exp2420 motes, Exp1101 mote, Exp1120 motes, etc. COOJA is a highly useful tool for CONTIKI development that allows the users to develop and test their codes before going to hardware (Fig. 1).

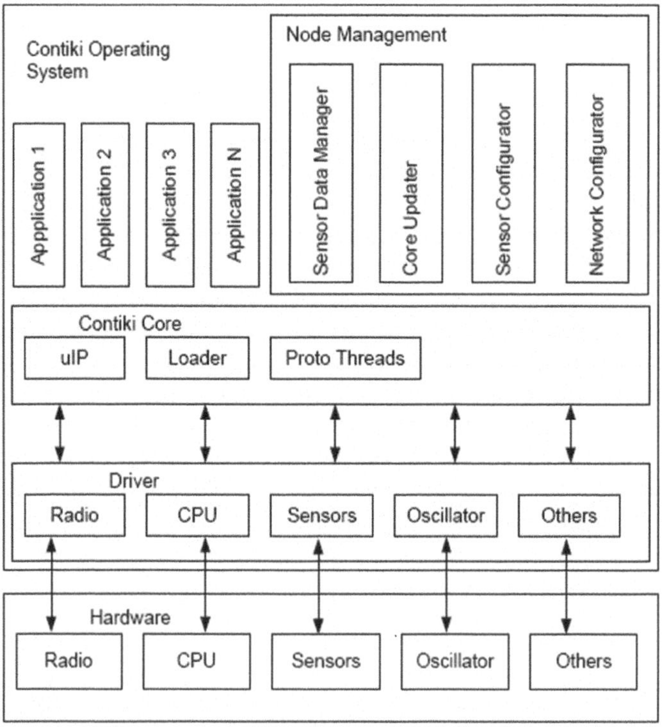

Fig. 1 CONTIKI architecture

Monitoring of sensor nodes can be done in CONTIKI by employing border router in the gateway unit to route the data between the device and the Internet. The objective is to collect the data from the various sensors that are employed in the sensing field and make the data visible to the remote person.

This paper is aimed to get the data of temperature and light sensors employed in five different locations in the sensing field. The data can be displayed on the Webpage for continuous monitoring. The proposed system not only supports local communication but also support global communication, which is one of the features of the Internet of Things. The IPv6 addressing is the additional feature for the system to communicate globally.

The rest of the paper is organized as Sect. 2 discusses the related work. Section 3 describes the block diagram of proposed work. Section 4 discusses about the results and conclusion in next section.

2 Related Work

Levis et al. [3] proposed tiny OS architecture which is suitable for wireless sensor networks. Tiny OS mote platform has been widely implementing in many applications. Tiny OS supports FIFO scheduling mechanism. In FIFO scheduling, some disadvantages such as tiny OS does not support real-time applications. If the application is real time, this is not a good choice.

Bhatti et al. [4] proposed Mantis OS. It supports round-robin scheduling policy with priority. While comparing CONTIKI OS or tiny OS with Mantis OS, the scheduler uses preemptive priority scheduling that supports real-time applications.

Cao et al. [5, 6] proposed lite OS, which is UNIX-like abstractions for wireless sensor network. The lite OS schedules the real-time tasks with round-robin algorithm. There are lot disadvantages of round-robin algorithm such as larger waiting and response time. So, a better scheduling algorithm is needed for real-time applications.

3 Methodology

3.1 Proposed Block Diagram

The block diagram consists of sensing nodes connected the gateway unit. In the gateway unit, the border router routes the data which is collected from the sensor units to the Internet. The border router acts like a connection between the sensing field and the Internet. The sensor network consists of a group of temperature and light sensors that are placed in distinct locations to measure the temperature and light intensity at that position. This paper employs IPv6 addressing for each sensor devices (Fig. 2).

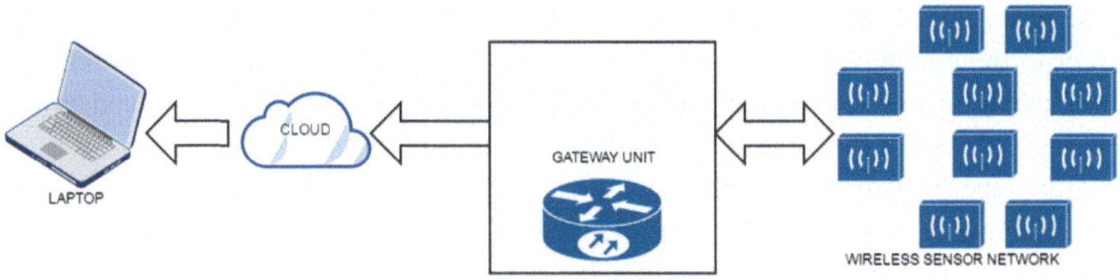

Fig. 2 Block diagram of proposed work

3.2 *Border Router*

The main motive of border router is to connect one type of network to another network. The sensor network employed in this paper is RPL network. The external network is Internet or cloud, which provides a Web access for monitoring (Fig. 3).

3.3 *CoAP Protocol*

The constrained application protocol (CoAP) is an open standard protocol used for constrained environments. The constrained nodes and network used CoAP to

Fig. 3 RPL network employed over sensor nodes

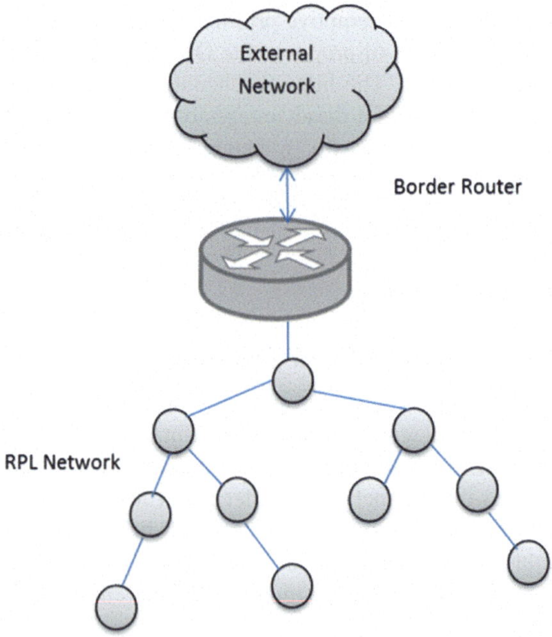

transfer data. CoAP uses UDP because of congestion in TCP. In IoT, lot of applications such as home automation and industrial surveillance use CoAP for transferring Web-based messages.

3.4 *RPL Protocol*

RPL is the IPv6 routing protocol for low power and lossy networks. The constraints for low power and lossy network (LLN) routers operate on memory, energy, and processing power. The RPL protocol works potentially up to thousands of nodes deployed in the network. In RPL, the traffic pattern will not be only in point-to-point communication but also it provides multipoint to point or point to multipoint. The unstable and lossy links provide low data rates and low packet delivery rates.

4 Results and Discussion

For simulation purpose, five nodes are taken for monitoring the conditions of the industry. Each sensing unit consists of light and temperature sensor attached to the microcontroller unit to get the data for monitoring.

Figure 4 shows simulation window consists of five sensors and a border router. The border router gets the data from the sensing units and displays on a Webpage.

Figure 5 shows the network consists of nodes and a border router that are interconnected by radio interference.

Figure 6 shows radio messages that are transmitting between sensor nodes and send it back to the gateway unit. The radio messages interface consists of a timestamp of messages transmitting from particular nodes to the group of nodes for transmitting the data.

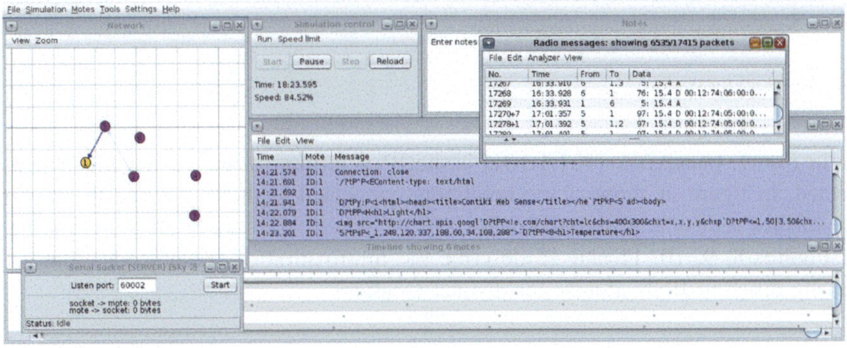

Fig. 4 COOJA simulator

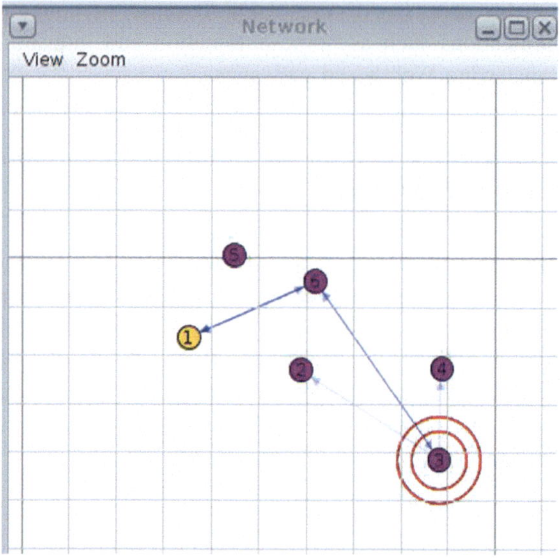

Fig. 5 Radio interference among the nodes

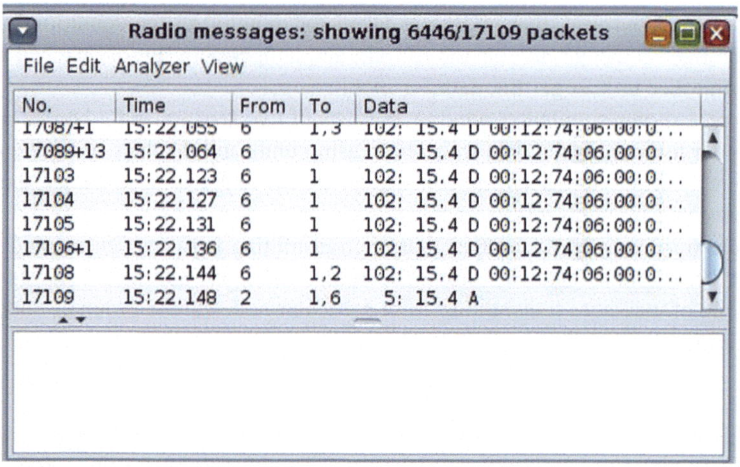

Fig. 6 Radio messages

In Fig. 7, [aaaa::212:7401:1:101] is the IPv6 format for the border router connected. In figure, it shows three neighbors connected to the router and the router establishes routes as mentioned below.

Figures 8, 9, 10, 11 and 12 describe the information of sensor nodes that are having different IPv6 address for distinct locations.

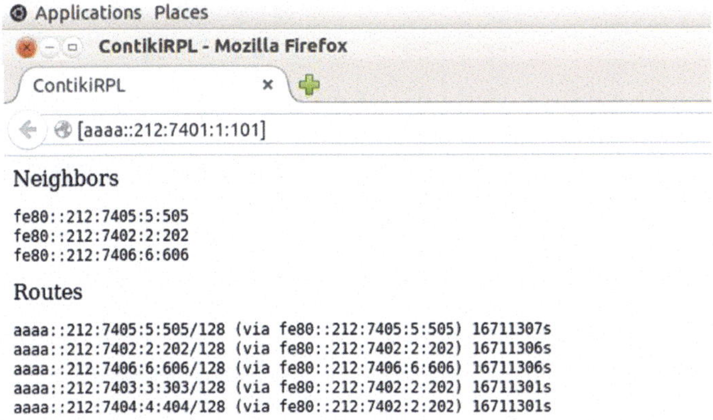

Fig. 7 Border router node

Fig. 8 NODE 2 information

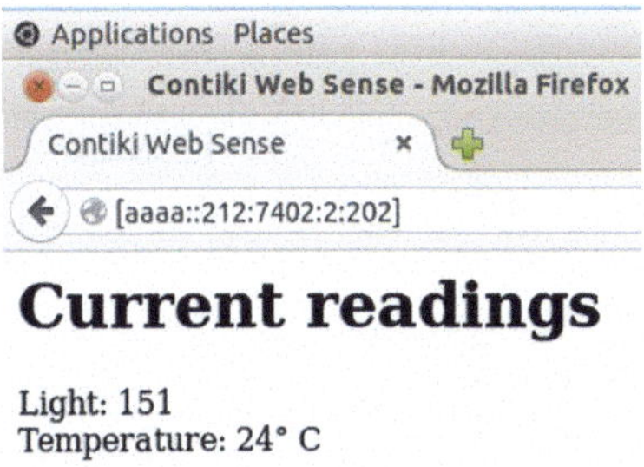

Fig. 9 NODE 3 information

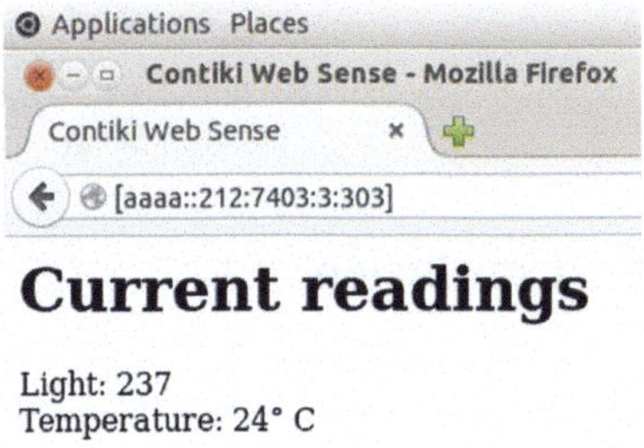

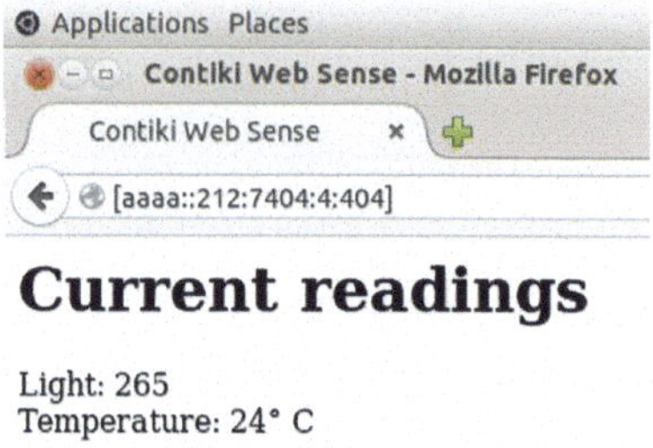

Fig. 10 NODE 4 information

Fig. 11 NODE 5 information

Fig. 12 NODE 6 information

5 Conclusion

This paper presents a periodic monitoring of sensor nodes employed in the remote area. The data gathered from the sensor nodes are monitored in a Webpage. Each node has distinct IPv6 addresses for communicating via IPv6 protocol. Further simulation reveals that each node is continuously monitoring in the Webpage by CONTIKI OS. N number of nodes can be monitored simultaneously by using the methodology employed in this paper.

Future work includes that the tasks from the Webpage can chat with sensor nodes in remote location by 6LowPAN adaptation layer.

Acknowledgements This research article is fully supported by VIT University. The authors would like to thank the reviewers for comments that improve the manuscript. I would like to thank my co-author for assisting the paper to complete.

References

1. Dunkels A, Gronvall B, Voigt T (2004) Contiki a lightweight and flexible operating system for tiny networked sensors. In: Proceedings of the 9th annual IEEE international conference on local computer networks, Washington, DC, USA, pp 455–462, Oct 2004
2. Osterlind F, Dunkels A, Eriksson J, Finne N, Voigt T (2006) Cross level sensor network simulation with COOJA. In: Proceedings of the 31st IEEE conference on local computer networks (LCN), Tampa, FL, USA, 14–16 Nov 2006
3. Levis P, Madden S, Polastre J, Szewczyk R, Whitehouse K, Woo A, Gay D, Hill J, Welsh M, Brewer E, Culler D (2005) Tinyos: an operating system for sensor networks
4. Bhatti S, Carlson J, Dai H, Deng J, Rose J, Sheth A, Shucker B, Gruenwald C, Torgerson HR (2005) Mantis OS: an embedded multithreaded operating system for wireless micro sensor platforms. Mobile Netw Appl 563–579
5. Cao Q, Abdelzaher T, Stankovic J, He T (2008) The LiteOS operating system: towards Unix like abstraction for wireless sensor networks. In: Proceedings of the 7th international conference on information processing in sensor networks IP, St. Louis, MO, USA, pp 22–24
6. LiteOS (2011) LiteOS [cit.2012-05-28]. Available from http://www.liteos.net

Analysis on LTE/Wi-Fi Data Offloading in Hetnets

C. Prasanth and S. Subashini

Abstract Due to extensive usage of smartphone, laptop, tablet, etc., there is an increase in volume of mobile data traffic. So, the best remedy to tackle with mobile data traffic issue is Wi-Fi offloading. It is the cost-effective approach that offloads mobile traffic in the existing Wi-Fi networks. Mobile users obtain data through Wi-Fi instead of cellular network; hence, it is the efficient technique to improve the spectrum efficiency and reduce the cellular network congestion. Wi-Fi access points have the advantage on amount of offload data and also provide quality of service in offloading data. To provide fairness between LTE and Wi-Fi, load balance is the technique addressed here. The load balancing problem can be solved by sharing traffic between adjacent cells. This work aims to provide an efficient dynamic load balance algorithm for offloading the data, and the simulation is done using MATLAB and NS2 software.

1 Introduction

Due to the increase in demand of mobile data usage by subscribers, it causes traffic overloading. Especially, during the peak hour, breaks in the user calls occur due to the insufficient network bandwidth. By 2016, in accordance with Cisco Visual Networking, 51% of IP load will be served by the wireless networks. The percentage of mobile data served by cellular network was 10%. Operators are considering both the technologies licensed (3GPP LTE) and unlicensed (Wi-Fi) to fulfill the demand. Wi-Fi networks are more capable than 4G networks to meet the increasing spectrum demand. Wi-Fi provides 680 MHz of new spectrum to operators. Recently, cellular operators rely on Wi-Fi offloading due to higher data rates,

---

C. Prasanth · S. Subashini (✉)
School of Electronics Engineering (SENSE), VIT University,
Chennai Campus, Chennai, Tamil Nadu, India
e-mail: subashini.s@vit.ac.in

C. Prasanth
e-mail: prasanth.c2015@vit.ac.in

A. M. Zungeru et al. (eds.), *Wireless Communication Networks and Internet of Things*, Lecture Notes in Electrical Engineering 493,
https://doi.org/10.1007/978-981-10-8663-2_9

low cost, etc. In the present scenario, 50% of the traffic related to cellular data is actively offloaded through spectrum that is unlicensed. Wi-Fi offloading is one of the implementations of using small cell technologies to provide data services to cellular users in a more efficient and economic manner.

2 Proposed Work

In this work, we developed an effective algorithm for dynamic load balance and offloading between LTE to Wi-Fi and vice versa.

2.1 *Dynamic Load Balance*

Initially, user equipment (UE) is connected to LTE network. If in the network any access point (AP) is present then, there will be exchange of beacons between UE and AP. Further AP will check for the maximum load (i.e., data size) in a region, if the load is greater than the threshold value (*B*th), then AP will reply with a negative beacon to the UE. This means AP is not ready to accept that particular UE in the network. However, If the load is less then threshold value, then AP will check for the received signal strength (RSS), so that there is less chance of congestion. Now AP will send a positive beacon to the UE stating that it is ready for the connection. If it has good signal strength it is offloaded, else remains in same channel (Fig. 1).

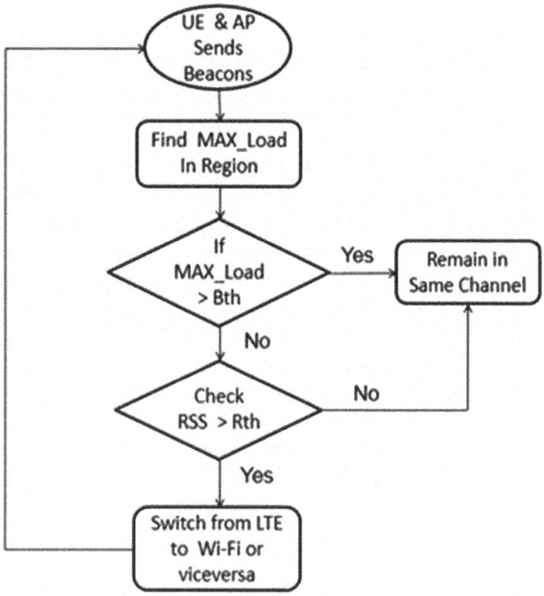

Fig. 1 Dynamic load balance

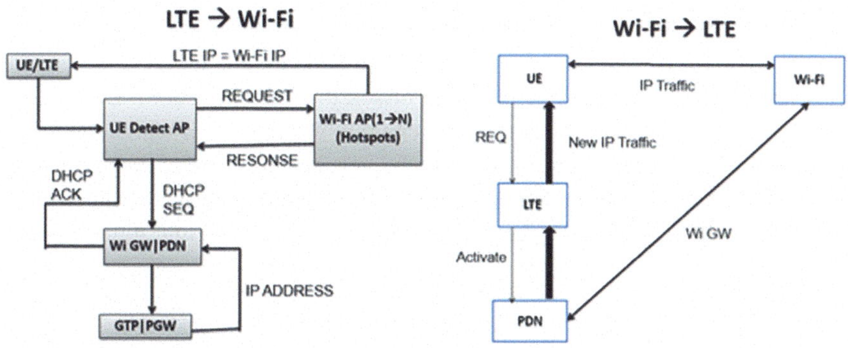

Fig. 2 Block diagram for switching between networks

2.2 *LTE to Wi-Fi Offload*

User is connected to LTE; it sends a beacon (Request and Response) to find a nearest AP with maximum signal strength. LTE works in all seven layers, whereas Wi-Fi works in lower three layers. UE sends a DHCP sequence to Wi-Fi gateway (Wi-GW), and it in turn sends a request to GPRS tunneling protocol (GTP) equipped with packet gateway (PGW) for IP address matching (i.e., to maintain an IP traffic if offloaded). Finally, Wi-GW sends a DHCP acknowledgement to UE. This results in offloading of traffic from LTE to Wi-Fi AP (Fig. 2).

2.3 *Wi-Fi to LTE Offload*

Consider an IP traffic flows from Wi-Fi to UE. When Wi-Fi AP channel is congested, Wi-Fi AP sends beacons to LTE via UE, if LTE channel is free, then LTE will activate the packet data network (PDN). New traffic will flows through Wi-GW for IP matching between Wi-Fi and LTE if data is offloaded. This results in formation of new IP traffic will flow from LTE to UE.

3 Mathematical Analysis

For mathematical analysis, some assumptions are made.

Assumption 1: To measure arrival rate of user, Poisson distribution process is used.
Assumption 2 (Independence assumption): Arrival rate of packets from BS or from AP is independent and identically distributed.
Assumption 3 (Round-robin scheduling): To bring fairness among all users present in different regions, the entire area is split into four regions.

3.1 To Find Maximum Load in Region

For determination of arrival rate of user, Poisson distribution is used. Let us consider R_n^{k+1} is the total number of users in region n, where n is the region identifier. R_n^k is the initial number of users, N_{k+1} is the new arrival of users. For switching between LTE to Wi-Fi or vice versa, a threshold bandwidth is set as B_t. α_n is term to determine number of users served or offloaded. B is the bandwidth multiplication factor.

For region n:

$$R_n^{k+1} = \begin{cases} R_n^k * B + N_{k+1} * B - \alpha_1; & R_n^k > 0 \\ N_{k+1} * B; & R_n^k = 0 \end{cases} \tag{1}$$

where $\alpha_n = \min\{R_n^k * B, B_t\}$,

Here, we take $n = 4$; i.e., the entire area is split into four regions. Substitute $n = 1, 2, 3, 4$ in the above equation, respectively.

From the above equations, we are estimating the region to be offloaded based on the maximum number of users present in the particular region.

$$\beta_1 = \max\{R_1^{k+1}, R_2^{k+1}\} \tag{2}$$

$$\beta_2 = \max\{R_3^{k+1}, R_4^{k+1}\} \tag{3}$$

$$\gamma = \max\{\beta_1, \beta_2\} \tag{4}$$

Here, γ determines the region to be offloaded.

3.2 System Model and Problem Definition

Let us model this scenario in M/M/1 queuing system. We assume a simple wireless network consisting of one mobile base station (BS) and Wi-Fi access point (AP). Each transceiver has a buffer of infinite capacity to incoming packets. Packets arrive from BS and AP with rate of A_m and A_w. The average departure of packets from each queue is D_m and D_w. We assume that time duration taken for packet to successfully transmit from BS and AP is exponentially distributed as $1/D_m$ and $1/D_w$. This time duration depends on channel condition between transmitter and receiver and packet length (Fig. 3).

Consider, if $A_m > A_w$ and $D_m < D_w$, then queue of BS is busy than AP (i.e., chance for congestion in BS). Then to release congestion of the queue, BS can offload part of packets from cellular traffic to Wi-Fi AP with probability of $(1 - P)$.

Fig. 3 System model

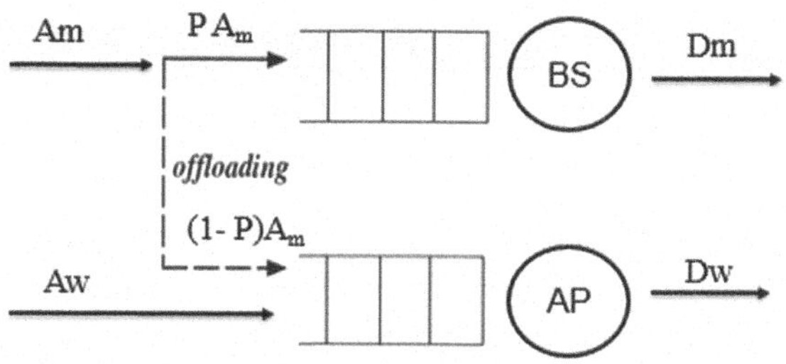

When $A_m < A_w$ and $D_m < D_w$, the queue of AP is busy than BS (i.e., AP is congested). Then to free the congestion of queue from AP, AP is offloaded a part of packets to BS with probability of P; here, P is the offload probability.

The stability of queue is determined by arrival and departure rates as A and D. If $A < D$, then queue is stable, and if $A > D$, then queue is unstable. In proposed offload scheme, the probability of packet arrival from queue is decomposed as PA_m to BS and $(1 - P)\,A_m$ to AP.

Let $A_{m,e}$ and $A_{w,e}$ be the effective arrival rates of queue from BS and AP and they are given as

$$A_{m,e} = PA_m \tag{8}$$

$$A_{w,e} = A_w + (1 - P)A_m \tag{9}$$

For stability of each queue, the offload probability should satisfy the below condition as $A_{m,e} < D_m$ and $A_{w,e} < D_w$, respectively.

4 Simulation Results

4.1 MATLAB Results

We simulated an actual scenario using MATLAB software. The users are arrived randomly (Poisson distribution) across the region and average arrival time of user is 100 s. The packet length of each user is 10 Mb. Here, we use four AP (i.e. $M = 4$). The simulation is run for two time slots. Threshold bandwidth is set as 50 MHz. By round-robin scheduling, data size is calculated in all 4 regions and offloaded according to the load balance condition (Fig. 4; Table 1).

Workspace

Name ▲	Value	Min	Max
A	50	50	50
A2	41	41	41
ArrivalTime	[10077,10022,10014,1...	10014	10154
AverageArrivalTime	[100,100,100,100]	100	100
AveragePacketLen...	50	50	50
B	50	50	50
B2	45	45	45
Bt	[50,50,50,50]	50	50
CurDataSize	[910,880,950,840]	840	950
Data_Threshold	1000	1000	1000
M	4	4	4
Noffload1	50	50	50
Noffload2	45	45	45
Nuser	<4x500 double>	0	95
PacketLength	[10,10,10,10]	10	10
R1	91	91	91
R2	88	88	88
R3	95	95	95
R4	84	84	84
SlotTime	2	2	2
TotalTime	10000	10000	10000
TotalTimeInSecon...	20000	20000	20000
alpha1	41	41	41
alpha2	38	38	38
alpha3	45	45	45
alpha4	34	34	34
i	4	4	4
j	4	4	4
k	4	4	4
n	4	4	4
r11	50	50	50
r12	41	41	41
r21	50	50	50
r22	38	38	38
r31	50	50	50
r32	45	45	45
r41	50	50	50
r42	34	34	34
t	10000	10000	10000
threshold	50	50	50

Fig. 4 Number of users offloaded based on γ value

Table 1 Users benefited by offloading

Users offloaded	Total number of users	Time interval	Users served by offloading
Region 1	91	50	41
Region 2	88	50	38
Region 3	95	50	45
Region 4	84	50	34

4.2 NS2 Results

We simulated an actual scenario using NS2 for visualization purpose; here, we use DSDV protocol for routing and propagation model as two ray ground. The switching of network will be on the basis of flow id assigned to nodes from different region. By using grep command, we determine MAX flow id such that switching of traffic between networks happens (Fig. 5).

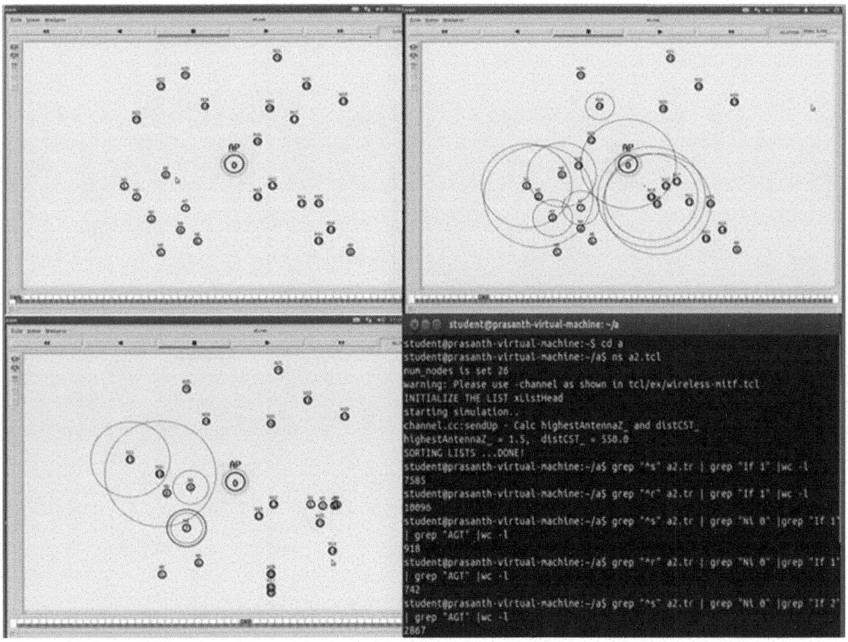

Fig. 5 Users offloaded based on DLB algorithm

5 Conclusion and Further Work

In this work, a dynamic load balance algorithm which is proposed here is the efficient technique for switching from LTE to Wi-Fi and vice versa, based on signal strength and channel utilization for heterogeneous networks. To justify the mathematical analysis, simulation of heavy load scenario is done using MATLAB and NS2 software tool. As a future work, we try to minimize the delay during switching.

Contention-Based CSI Feedback Mechanisms in MU-MIMO WLANs: A Survey

D. Srinivasa Rao and V. Berlin Hency

Abstract In this paper, the significance of the CSI feedback acquisition at the WLAN AP is highlighted. CSI is the channel state information obtained from the physical layer to support the throughput enhancement in the upper layer. However, this information should be optimized in order to mitigate the feedback overhead with the increasing number of users. Here, a survey of existing CSI feedback algorithms with the current 802.11 WLAN standards is done by considering factors like overhead, delay, and throughput performance. We also discussed the user selection and scheduling in MU-MIMO-based WLANs. The scheduling is usually performed with the help of feedback CSI from selected users. However, acquiring CSI from all user STAs results in large overhead and grows linearly with the channel sounding interval and with the increased number of users. The current literature is more focused on reducing CSI overhead. In order to fully realize the benefit of MU-MIMO and guarantee the required QoS, it is important to acquire updated CSI from all the users. Hence, there exists a trade-off between efficiency of scheduler and CSI overhead. Generally, the AP limits the number of users based on feedback CSI. The best user CSI and suitable channels are needed to be obtained before the user is scheduled.

Keywords MU-MIMO · WLAN · CSI · ZFBF

D. Srinivasa Rao
Department of ECE, GMR Institute of Technology, Rajam, Andhra Pradesh, India
e-mail: srinivasa.dasari@gmail.com

V. Berlin Hency (✉)
School of Electronics Engineering, VIT University, Chennai, Tamil Nadu, India
e-mail: berlinhency.v@gmail.com; berlinhency.victor@vit.ac.in

A. M. Zungeru et al. (eds.), *Wireless Communication Networks and Internet of Things*, Lecture Notes in Electrical Engineering 493,
https://doi.org/10.1007/978-981-10-8663-2_10

1 Introduction

Multi-user multiple-input and multiple-output (MU-MIMO) is considered as the key technique for the future wireless local area networks (WLANs). Current 802.11ac [1] allows multiple transmissions to the users with the help of spatial division multiplexing. This attains improved spectral efficiency over single-user MIMO communication [2]. The MU-MIMO [3] systems require full CSI information at the transmitter in order to perform well over SU-MIMO systems. This necessitates the acquisition of CSI from all the competing users. Most of the useful time is spent on transmission of feedback requests. With the increase in the number of transmit antennas and number of users, the CSI overhead [4] increases as shown in Fig. 1. This leads to performance degradation of next-generation WLANs. To realize the benefit of MU-MIMO, it is highly essential to develop efficient CSI feedback mechanisms for these WLANs.

In the downlink MU-MIMO WLANs, the fundamental problem is to handle multiple requests from users at the same time. This problem can be split up into two phases. In the first phase, the access point has to send the channel probe packets with forward link, and in next phase, best users should feedback the CSI in reverse direction. After collecting the CSI from all the required users, the MU-MIMO transmission is scheduled. During the second phase, the best users have to follow the contention mechanism to successfully transmit the CSI information to the AP. These are termed as contention rounds. If there are M transmit antennas and N number of single antenna users, then there will be a maximum of $M - 1$ contention rounds. In each round, the user whose channel satisfies the threshold requirements contends for resources. Any two users who might pretend their CSI is best may transmit the CSI at the same time, and this leads to collisions. If the number of users increases, the overhead increases and the number of collisions also increases. This is the main motivation behind the survey. The remaining paper is organized as follows. Section 2 describes the CSI feedback mechanism of the current 802.11ac WLAN. In Sect. 3, a brief survey of the state-of-the-art CSI feedback mechanisms is given. Finally, Sect. 4 concludes the paper.

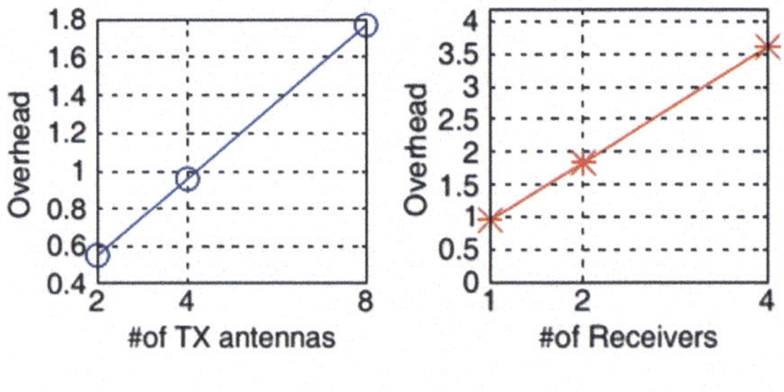

Fig. 1 CSI feedback overhead increases with the number of transmit antennas and receivers [4]

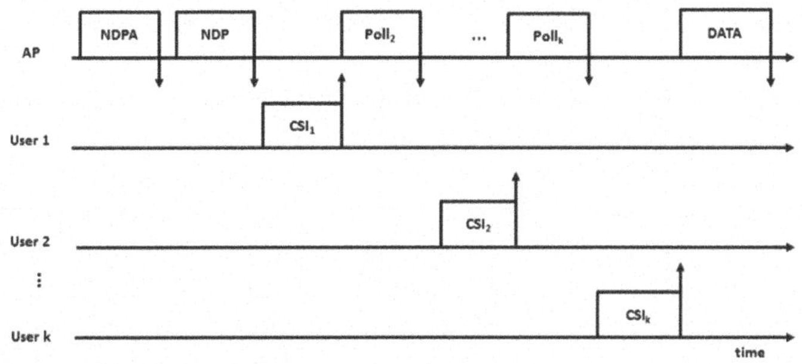

Fig. 2 CSI feedback mechanism [6]

2 802.11ac CSI Feedback

The 802.11ac MIMO networks [5] follow polling procedure to obtain the CSI feedback from the intended users. The CSI feedback operation for 802.11ac networks is shown in Fig. 2.

First, the AP will transmit null data packet announcement (NDPA) packets to initiate the procedure. Null data packet (NDP) acts as preamble which contains training sequence. Upon receiving the NDP packet, each user calculates its channel state vector and feedbacks to the transmitter. In the example shown, user 1 transmits its CSI to the AP. Then, remaining k users will feedback their CSI to the access point. After collecting all users CSI, the AP prepares the precoding matrix and starts MU-MIMO transmission. The ordering of users is conveyed in NDPA packet. Hence, 802.11ac [6] needs to collect CSI from all the users before initiating scheduling. Unfortunately, as the number of users increases, the overhead incurred in transmission increases. Sometimes the data have to be sent at low rates. So, 802.11ac indulges huge overhead with the increase in the number of users.

To reduce the feedback overhead, 802.11ac goes for quantization up to 8 bits and permits maximum of four adjacent OFDM frequency bins to form CSI. Even after such compression, the user station overhead still ranges from 100 to 800 bytes and increases along with the number of envisioned users. Instead, CSI report can be sent very often mentioned in [7]. But to guarantee accuracy, the feedback interval must be chosen smaller than the channel coherence time. In indoor or static environments, in order to ensure accuracy, the user's feedback period needs to be shorter than 15 ms.

3 Survey of CSI Feedback Mechanisms

In this section, a review of some state-of-the-art CSI feedback mechanisms is provided. The CSI feedback can be done explicitly or implicitly [8]. In explicit feedback, the user stations compute the individual CSI estimates and send to base station. While in implicit feedback, the access point or base station is responsible for user selection and scheduling. In [9], the author compared implicit and explicit feedback mechanisms based on overhead requirements and packet error rate performance. In MU-MIMO WLANs, the major is to handle multiple requests at the same time. As the number of users increases, the overhead associated with other stations also increases. This results in reduction of overall throughput of the system. Many authors emphasized the problem of maximizing sum rate. Most the researchers considered this as a non-convex optimization problem. Thus, achieving high sum rate or throughput is an optimization problem in multi-user wireless networks. Due to the high computational complexity in encoding and decoding, optimal algorithms are difficult to implement in practice. Hence, in [10], the author proposed a suboptimal algorithm called semiorthogonal user selection (SUS) and shown that it can achieve asymptotic sum capacity as the number of users goes to infinity. In this paper, the author employed zero-forcing beamforming strategy to cancel the mutual interference among the users. The relation among the users is studied to check the possibility of grouping and enhance the throughput. Fairness is studied using round-robin ZFBF and proportional fair ZFBF scheduling schemes. The complexity is less, because ZFBF strategy is simple transmit precoding scheme.

A greedy user selection algorithm (GUSS) is proposed in [11] to eliminate the complexity involved in the zero-forcing selection and achieves the upper bound of sum rate with low complexity. Here, the author considered throughput and complexity measures to evaluate the performance of the algorithm. SUS and GUSS scheme provides high throughput for multi-antenna base station with multi-users simultaneously. Recently, an orthogonality probing-based user selection (OPUS) mechanism has been proposed in [12] that does not require CSI from all the users. In MU-MIMO network, 'M' be the number of antennas at the access point and 'N' be the number of users. OPUS requires up to M rounds of CSI feedback, and the orthogonality among the users is evaluated using novel probing mechanism. Further, a distributed contention-based feedback mechanism is proposed that singles out the best user among the contending users. The probing mechanism along with the contention feedback scheme makes it suitable for downlink MU-MIMO WLAN. Here, the author evaluated the performance of OPUS algorithm by comparing throughput and fairness with traditional user selection schemes.

The feedback contention mechanism of OPUS algorithm is explained in Fig. 3. The unselected users will compute and quantize the CSI feedback into N bits. There will be N contention stages based on number of quantizing bits. Each user has to undergo N-bit contention stages. User with '1' as corresponding bit sends short energy pulse, and user with '0' bit listens to the channel. User with '0' as corresponding bit stop contending if they find the channel is busy. In this way, each will go through N-bit contention stages and finally, one user will be the contention. OPUS has contention overhead fixed up to $3N$ slots. OPUS maintains low collision probability by keeping upper and lower bounds on SINR values. However, OPUS suffers from the hidden node problem as each can overhear other. OPUS enhances downlink throughput and guarantees fairness among the users.

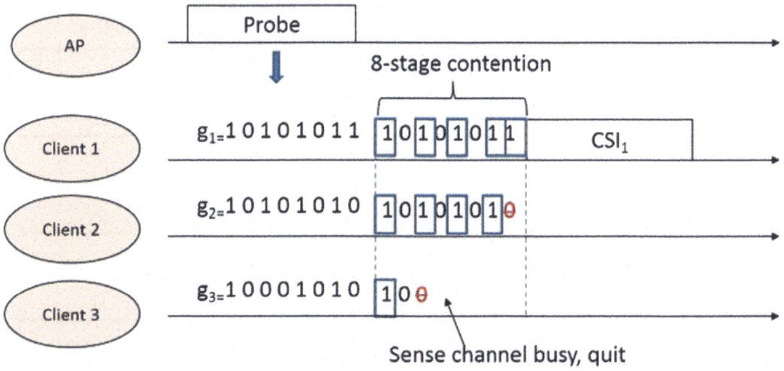

Fig. 3 OPUS CSI feedback contention mechanism [12]

In [13], the author presented a novel downlink MU-MIMO sounding protocol called MUTE which reduces the overhead linked with sounding and maximizes the user selection performance. The proposed design is evaluated and implemented. The author particularly deals with the problem of decoupling the sounding and user selection procedure in ZFBF transmission. This facilitates the AP to select the users independently based on the user channel characteristics, i.e., whether to sound a particular user or not. This in turn results in reduced overhead with user sounding by identifying the existence of users with stable channel conditions and providing sufficient information to the AP about the user channel conditions. Then, using this information, the AP can choose the group of users that maximizes an objective function such as achievable rate or a fairness criterion as example. This differs from the existing MU-MIMO techniques where the group of users who are sounded is the same as the set of users that are to be assisted next.

In [14], the author proposed a novel user selection algorithm 802.11ac+ to increase the system capacity. 802.11ac+ comprises of a novel MU-MIMO MAC protocol and delay transmission approach to transmit the CSI feedback. The first user will be selected randomly and requests CSI feedback. Based on the CSI feedback obtained from user 1, access point will prepare a channel hint and broadcasts it to remaining users. Using the channel hint, each user computes its channel state vector and contends for channel in distributed manner. The contention feedback mechanism is shown in Fig. 4.

Feedback contention arises when two users transmit the CSI feedback at the same time. 802.11ac+ avoids contention by using delayed feedback approach. In this approach, the users are allowed to transmit their feedback only in specific time slots. Each slot has fixed some threshold, and the user whose CSI satisfies the threshold value wins the contention and announced as winner of the contention round. Likewise, there will be maximum of '*M*' contention rounds for *M* antenna access point. The performance of the contention mechanism depends on how well the timeout threshold and slot thresholds are fixed. The author also proposed two fair scheduling protocols, namely round-robin scheduler and proportional fair scheduler to solve the fairness problem among the users. The throughput obtained

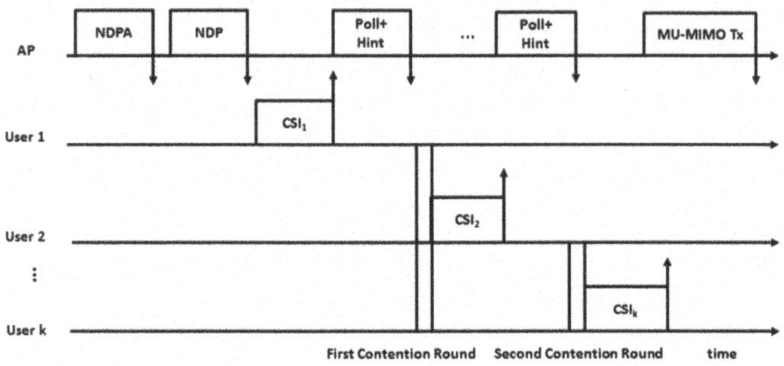

Fig. 4 802.11ac+ MU-MIMO transmission [14]

using this scheme is shown better compared to SUS and 802.11ac protocol. However, the author has not clearly addressed the complexity and overhead issue in the proposed scheme. In addition to the above downlink MU-MIMO WLAN contention mechanisms, we also discuss two uplink contention mechanisms which can be extended to downlink and may yield better results.

In [15], the author proposed a novel orthogonality evaluation mechanism that enables each user to obtain its own CSI. This algorithm is named as signpost. Signpost also realizes a 2D-prioritized contention mechanism to choose the best user efficiently by using both time and frequency domain resources. Signpost is a scalable user selection algorithm that is suited uplink MU-MIMO WLAN transmission. In this protocol, for each contention round, arbitrary probing directions are transmitted as channel hint to the user stations. The user stations check the orthogonality using these arbitrary directions and contend for the channel without sending the feedback to access point. Hence, with zero CSI overhead, the user competes for the resources. The contention mechanism is shown in Fig. 5.

Figure 5 shows an example of signpost contention mechanism. Here, $G_{i,j}$ is quantized preference metric in *j*th signpost direction corresponding user *i*. The alignment for three users and four subcarriers are mentioned in the above diagram. Here, user 1 wins the contention on first direction and user will win on second direction. Each alignment metric is mapped to subcarrier index and time slot. Hence, signpost uses both time and frequency domain resources to avoid the contention during CSI feedback phase. This is termed as two-dimensional feedback

Fig. 5 Signpost contention mechanism [15]

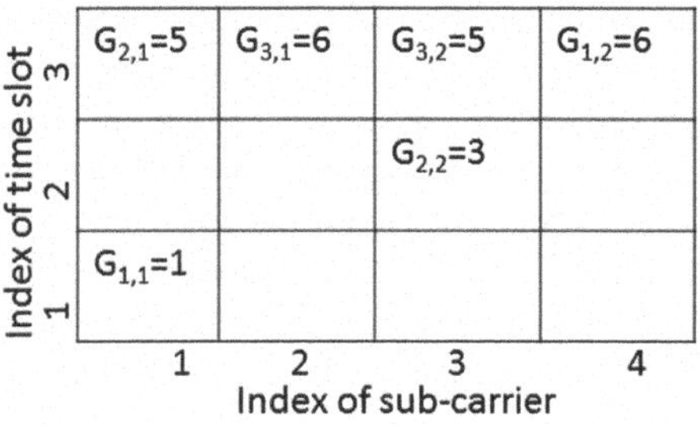

contention mechanism. The author evaluated the performance of the protocol by considering throughput and overhead issues. Another recent uplink MU-MIMO WLAN protocol optimal user selection (OUS) is proposed in [16]. OUS takes throughput and fairness into consideration and formulates the complex scheduling problem. OUS also considers correlation among the users and provides throughput fairness solution to the user selection problem. It studies the impact of grouping the users on throughput and fairness. OUS is an uplink scheduling scheme and can be extended to downlink MU-MIMO WLANs. The scheduling operation of OUS is shown in Fig. 6. The entire scheduling period of OUS is divided into three stages. They are CSI estimation, scheduling, and transmission. Here, an AP and several users are considered for transmission. The AP that acts as central controller takes decisions on user selection-based channel correlation. The sequence of operations is as follows. First, the AP broadcasts signals to user stations to announce uplink transmission. Then, the user responds to the AP with their individual CSIs. After this, the AP calculates the SNRs and channel correlation '*r*' between the users. Up to this step, OUS performs usual procedure. These values are provided as input to the OUS greedy selection algorithm which is the key component of the scheme. This algorithm calculates and outputs the set of concurrent users for each transmission slot. Finally, the AP informs users about the data rates and allows them to transmit concurrently.

The performance metrics of these CSI feedback schemes are provided in Table 1. The parameters considered are throughput, fairness, overhead, and complexity. It presents the key design issues considered in the state-of-the-art downlink and uplink CSI feedback mechanisms.

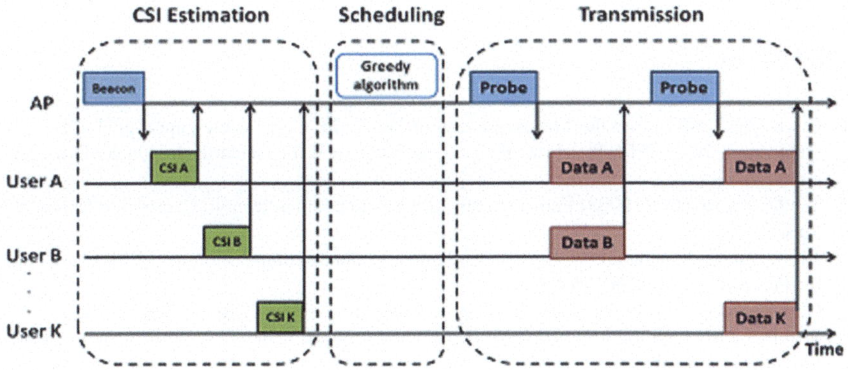

Fig. 6 Scheduling operation flow of OUS [16]

Table 1 Performance metrics of CSI feedback schemes

Scheme/Algorithm	Design issues considered			
	Throughput	Fairness	Overhead	Complexity
SUS [10]	✓	✓		✓
GUSS [11]	✓			
OPUS [12]	✓	✓	✓	
MUTE [13]	✓		✓	
802.11ac+ [14]	✓	✓		
Signpost [15]	✓		✓	
OUS [16]	✓	✓		

4 Conclusion

In this paper, a survey of contention-based CSI feedback schemes is done for MU-MIMO WLANs. User selection, scheduling, and feedback contention are considered as major problems in upcoming wireless local area networks. As the number of user stations increases, the overhead incurred during transmission increases. This leads to degradation of the overall throughput. This is considered as the motivation for this survey. Most recent CSI feedback mechanisms are studied with respect to performance issues such as throughput, fairness, overhead, and complexity. It is concluded that there is a need to develop efficient CSI feedback schemes to support the high physical data rates offered by the next-generation WLANs.

References

1. Cisco (2012) 802.11ac: the fifth generation of Wi-Fi. In: Cisco white paper, pp 1–25
2. Qualcomm (2012) IEEE 802.11ac: the next evolution of WiFi standards. In: Qualcomm white paper, pp 1–13
3. Tse D, Viswanath P (2005) Fundamentals of wireless communication. Cambridge University Press
4. Xie X, Zhang X, Sundaresan K (2013) Adaptive feedback compression for MIMO networks. In: Proceedings of ACM MobiCom
5. Gast M (2005) 802.11 wireless networks: the definitive guide. O'Reilly Media, Inc.
6. Wireless LAN medium access control (MAC) and physical layer (PHY) specifications. IEEE Std. 802.11ac Draft 3.0
7. Liao R, Bellalta B, Oliver M, Niu Z (2014) MU-MIMO MAC protocols for wireless local area networks: a survey. arXiv: 1404.1622v2
8. Gong MX, Perahia E (2010) Training protocols for multi-user MIMO wireless LANs. PIMRC, IEEE
9. Lou H, Ghosh M, Xia P, Olesen R (2013) A comparison of implicit and explicit channel feedback methods for MU-MIMO WLAN systems. IEEE
10. Yoo T, Goldsmith A (2006) On the optimality of multiantenna broadcast scheduling using zero-forcing beafforming. IEEE J Sel Areas Commun 24(3)

11. Huang S, Yin H, Wu J, Leung VCM (2013) User selection for multiuser MIMO downlink with zero-forcing beamforming. IEEE Trans Veh Technol 62(7)
12. Xie X, Zhang X (2014) Scalable user selection for MU-MIMO networks. In: Proceedings of IEEE INFOCOM 2014
13. Bejarano O, Magistretti E, Gurewitz O, Knightly EW (2014) MUTE: sounding inhibition for MU-MIMO WLANs. IEEE
14. Lee K, Kim C-K (2015) User scheduling for MU-MIMO transmission with active CSI feedback. EURASIP J Wirel Commun Netw 2015(1):112
15. Zhou A, Wei T, Zhang X, Liu M, Li Z (2015) Signpost: scalable MU-MIMO signaling with zero CSI feedback. ACM
16. Zhou Y, Zhou A, Liu M (2016) OUS: optimal user selection in MU-MIMO WLANs. IEEE

Synchronization Analysis of Quadratic Chaos-Based DSSS-OFDMA System with an Interceptional Attack

R. Priya and R. Kumar

Abstract Chaotic communication system is divergent from spread spectrum communication which is not as easy to find an accurate application as it can arise non-uniformity equations. The conventional chaos is appropriately worse for a forthcoming state as because it is higly sensitive to an early state under jamming. The conservative system of proposed quadratic chaos has been beneficial in many technological areas even in the presence of jamming. The quadratic chaos is easier to generate, and it is hard to detect by fraudulent users. Therefore, the quadratic chaotic sequence with orthogonality property has the ability to improve signal detection and also to suppress the multicarrier interference. In the proposed method of quadratic chaotic DSSS-OFDMA system for distinct values of increasing jammer power, the signal detection under jamming has been improved than existing chaos.

Keywords Jamming attack · Chaos-based communications · Synchronization Physical layer security

1 Introduction

It is crucial to examine the performance of wireless kind of applications against security attack in which jamming is precarious and may disrupt the communication [1]. A chaos may exhibit a variation to an early condition under interference attack. The prognostic close to a forthcoming state are progressively worse even in an initial state for existing chaos.

R. Priya (✉) · R. Kumar
Department of Electronics and Communication, SRM University, Kattankulathur, Tamil Nadu, India
e-mail: priya_r@srmuniv.edu.in

R. Kumar
e-mail: kumar.r@ktr.srmuniv.ac.in

A. M. Zungeru et al. (eds.), *Wireless Communication Networks and Internet of Things*, Lecture Notes in Electrical Engineering 493,
https://doi.org/10.1007/978-981-10-8663-2_11

The conventional system: It has morbid issue while assigning values to initial condition parameter as it persuaded to negative population sizes. Therefore, to depreciate the interference and complexity level, quadratic chaos is suitable way to suppress the non-uniformity in traditional chaos.

2 System Description

2.1 Quadratic Chaotic System

The pictorial representation of quadratic chaos DSSS-OFDMA system has been demonstrated in Fig. 1 in which this system makes use of orthogonal quadratic chaotic sequences to present a data bit. Thus, the transmitter spreads the source bits by quadratic chaotic sequence (q_i) from the *j*th bit [2]. The obtained sequence is entered through mapper. An efficient authentication-based Chebyshev polynomial map has the following function,

$$x_{k+1} = \cos\left(\omega \cos^{-1} x_k\right), \quad \text{for} -1 \leq x_k \leq 1 \tag{1}$$

The power of integer ω is 2. Therefore, the quadratic chaotic sequence for aimed system DSSS-OFDMA under jamming is represented as,

$$x_{n+1} = \sin^2(2^n \theta \pi) \tag{2}$$

where θ is considered as initial conditional parameter.

$$\theta = \frac{1}{\pi} \sin^{-1}\left((x_0)^{1/2}\right), \quad \text{for} -1 \leq x_0 \leq 1 \tag{3}$$

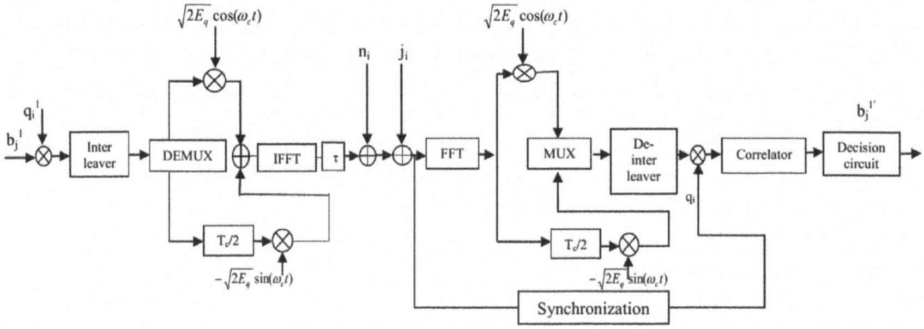

Fig. 1 Block schematic for quadratic chaos DSSS-OFDMA system

Using probability distribution for quadratic chaos sequence is considered as,

$$f_c(q_i) = \frac{1}{\pi\sqrt{2 - q_i^2}} \tag{4}$$

The quadratic chaotic sequence for the transmitted bits is shown as,

$$q_i = (x_n + 1) * 100/2 \tag{5}$$

The energy of quadratic chaotic sequence is expressed as,

$$E(q_i) = q_i^4 \int\limits_{-\infty}^{\infty} \frac{1}{\pi\sqrt{2 - q_i^2}} \mathrm{d}q_i \tag{6}$$

At the side of receiver, the evaluated result of quadratic chaotic sequence is then entered into correlator and decision circuit.

2.2 *Jammer*

The proposed system in which OFDMA system uses orthogonality sequences as it is highly resistant to jamming as compared with an efficiency in existing system of CDMA.

$$j(n) = j_I(n)\sqrt{2E_j}\cos(\omega_c t) - j_Q(n)\sqrt{2E_j}\sin(\omega_c t) \tag{7}$$

3 Synchronization of Quadratic Chaotic DSSS-OFDMA System

The synchronization is essentially used in order to examine the execution of the aimed system [3]. The correlator for the modern system is estimated as,

$$W = \sqrt{E_q}\sum_{i=1}^{2\beta} q_i' q_{(i-\tau)}' + \sqrt{E_n}\sum_{i=1}^{2\beta} n_i q_{(i-\tau)}' + \sqrt{E_j}\sum_{i=1}^{2\beta} j_i q_{(i-\tau)}' \tag{8}$$

where E_q is the chip energy of quadratic chaotic sequence, E_n is the noise energy, and E_j is the jammer energy [4]. Also, 2β is the chip sequence per E_b. The block schematic for synchronization of quadratic chaotic system is exposed in Fig. 2.

The received signal is then multiplied with quadratic chaotic sequences to retrieve the generated signal, and then, it is passed through correlator and square

Fig. 2 Synchronized proposed system with an interception

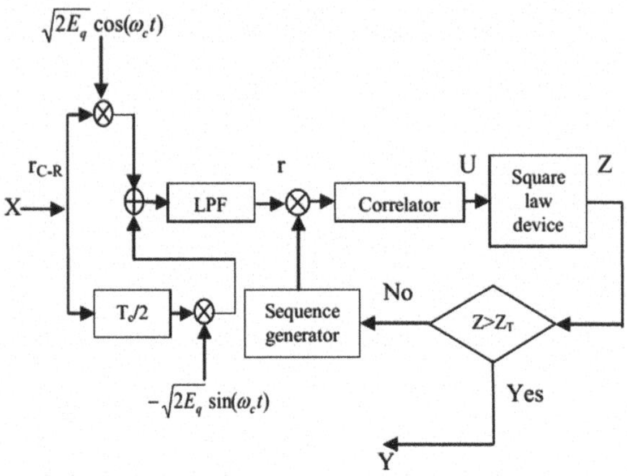

law device. Inorder to achieve synchronization, the outcome of square law device (Z) greater than the threshold value (Z_T).

4 Mathematical Analysis

4.1 Analysis of Quadratic Chaotic System

In general, the chaos system is considered as,

$$x_{n+1} = Ax_n(1 - x_n) \tag{9}$$

whereas '*A*' is in the range between 3.57 and 4. The quadratic chaotic system for the differential equation is assumed as,

$$\ddot{x} + b\dot{x} + kx^3 = A\sin\omega t \tag{10}$$

The variable can be specified as $\phi = \omega t$, and the quadratic chaos system is expected as,

$$\dot{x} = v, \quad \dot{v} = -bv - kx^3 + A\sin\phi \tag{11}$$

The simplest equation that can be evinced as,

$$\dot{x} = y, \quad \dot{y} = -x + yz, \quad \dot{z} = 1 - y^2 \tag{12}$$

The consistent system of quadratic chaotic for cubic system is deduced as,

$$\dddot{x} + x^2\dot{x} - A(1 - x^2)x = 0 \tag{13}$$

Using Euler method, the equation obtained for the transient-free system is determined as,

$$x_{n+1} = x_n + hv_n, \quad v_{n+1} = v_n - hx_n + 1 \tag{14}$$

4.2 False Alarm Under Jamming

This is contrary to the sensitivity as it does not contradict the received with generated signal. The correlator for specificity is resolved as,

$$U = \sqrt{E_q}\sum_{j=1}^{2\beta} q_i^0 q_{(i-\tau)}^0 + \sqrt{E_n}\sum_{j=1}^{2\beta} n_i q_{(i-\tau)}^0 + \sqrt{E_j}\sum_{j=1}^{2\beta} j_i q_{(i-\tau)}^0 \tag{15}$$

The mean value is considered as

$$E[U] = 0 \tag{16}$$

The variance for the above (16) is measured as,

$$\sigma^2 = E\left[U^2\right] - E^2[U] \tag{17}$$

$$\sigma^2 = E\left[\sqrt{E_q}\sum_{i=1}^{2\beta} q_i^0 q_{(i-\tau)}^0\right]^2 + E\left[\sqrt{E_n}\sum_{i=1}^{2\beta} n_i q_{(i-\tau)}^0\right]^2 + E\left[\sqrt{E_n}\sum_{i=1}^{2\beta} j_i q_{(i-\tau)}^0\right]^2 \tag{18}$$

$$\sigma^2 = E_q 2\beta + 2\beta n_o/2 + 2\beta n_j/2 \tag{19}$$

The Pareto distribution has been estimated as,

$$P_y = \frac{1}{\sigma(z)^2\sqrt{2\pi}} \mathrm{e}^{\left(-1/2\left(z/\sigma^2\right)\right)} \tag{20}$$

The false-positive rate with error function has been taken into consideration as,

$$P_F = 1 - \operatorname{erfc}\left[\sqrt{\frac{z_T}{2\sigma^2}}\right] \tag{21}$$

4.3 Signal Detection Under Jamming

The synchronization analysis has been performed for the signal detection by aligning the generated and received signal. The correlator for aligned signals is computed as,

$$U = \sqrt{E_q}\sum_{j=1}^{2\beta}(q_i^0)^2 + \sqrt{E_n}\sum_{j=1}^{2\beta}n_i q_i^0 + \sqrt{E_j}\sum_{j=1}^{2\beta}j_i q_i^0 \tag{22}$$

The average rate for (22) is revealed as,

$$E[U] = E\left[\sqrt{E_q}\sum_{j=1}^{2\beta}(q_i^0)^2 + \sqrt{E_n}\sum_{j=1}^{2\beta}n_i q_i^0 + \sqrt{E_j}\sum_{j=1}^{2\beta}j_i q_i^0\right] \tag{23}$$

where

$$E\left[\sqrt{E_q}\sum_{i=1}^{2\beta}(q_i)^2\right] = \sqrt{E_q}2\beta \tag{24}$$

$$E\left[\sqrt{E_n}\sum_{i=1}^{2\beta}n_i q_i^0\right] = 0 \tag{25}$$

$$E\left[\sqrt{E_j}\sum_{i=1}^{2\beta}j_i q_i^0\right] = 0 \tag{26}$$

The variance for the above expression is resolved as,

$$\sigma^2 = E_q\beta + 2\beta n_0/2 + 2\beta n_j/2 \tag{27}$$

The contiguous random distribution is estimated as,

$$P_y = \left[\frac{1}{\sigma\sqrt{2\pi z}}\exp\left[-\frac{[z+\lambda/\sigma^2]}{2}\right]\cosh\sqrt{\frac{z\lambda}{\sigma^4}}\right] \tag{28}$$

where $\lambda = E_q 4\beta^2$, the true positive rate from (28) is derived, and it is considered as,

$$P_D = 1 - \frac{1}{2}\left[\operatorname{erfc}\left(\sqrt{z_T/2\sigma^2} - \sqrt{\lambda/2\sigma^2}\right) + \operatorname{erfc}\left(\sqrt{z_T/2\sigma^2} + \sqrt{\lambda/2\sigma^2}\right)\right] \tag{29}$$

5 Systematic Results and Discussion

In accordance with the result of analytical scenario, a threshold value is considered as 200 in favor of signal detection. As this is counterfeit, the fixed point has been set in which variation can be comprehend by means of chip sequences per bit. The receiver characteristic curve is exposed in Fig. 3, for dissimilar values of SJR in which the chip sequence per bit energy is set as 300 [5]. The sensitivity has been highly improved for the proposed system of DSSS-OFDMA using quadratic chaos sequence with an interceptional attack. When decreasing the SJR, it is arduous to elucidate the signal as it leads to the error rate. As increasing SJR, it reaches the maximum probability under jamming.

To gain the efficiency of system by considering the boundary for divergent values of SJR, in which the sensitivity for the aimed system is enhanced than conventional chaos as viewable in Fig. 4.

Within that fixed point, the sensitivity is assessed by means of jammer power. After that considerable point, the communication system falls to error rate under interception. An error function for the analytical results has been related with simulated waveform.

The sensitivity based on SJR for dissimilar values of threshold has been exposed in Fig. 5. When SJR = 5 dB, the system performance is getting optimized with respect to the value of signal detection and threshold under jamming. The execution of the communication system falls to error rate for SJR = 12 dB. Therefore,

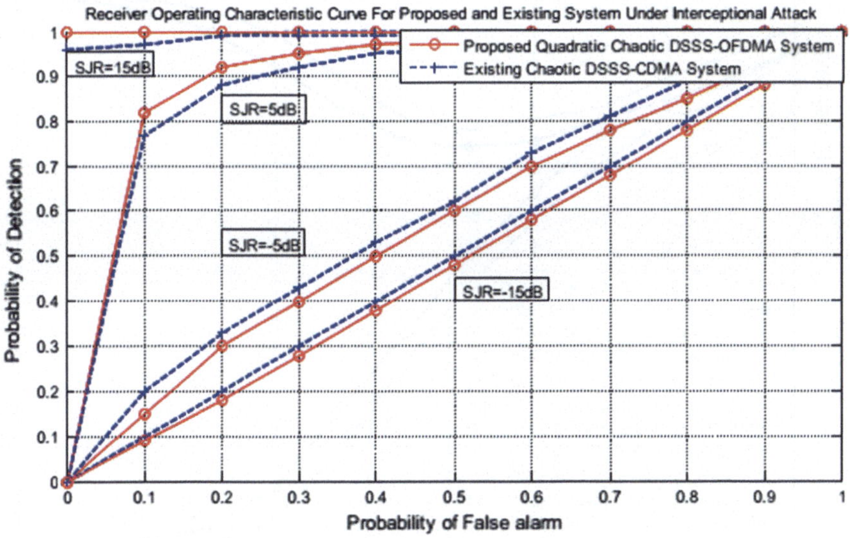

Fig. 3 Characteristic curve of received signal for the proposed and existing systems using dissimilar values of SJR with an interception

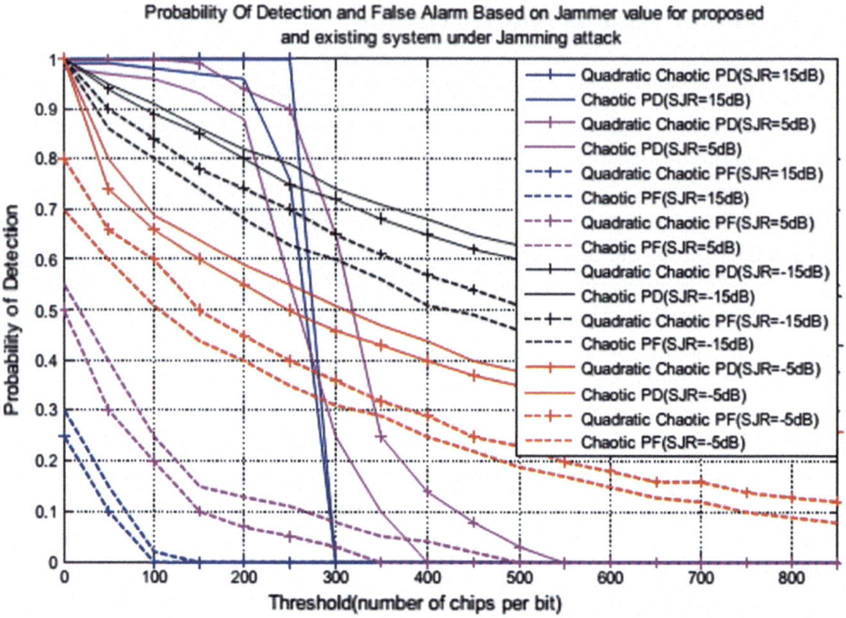

Fig. 4 Threshold-based signal detection system for proposed and existing systems with an interceptional attack

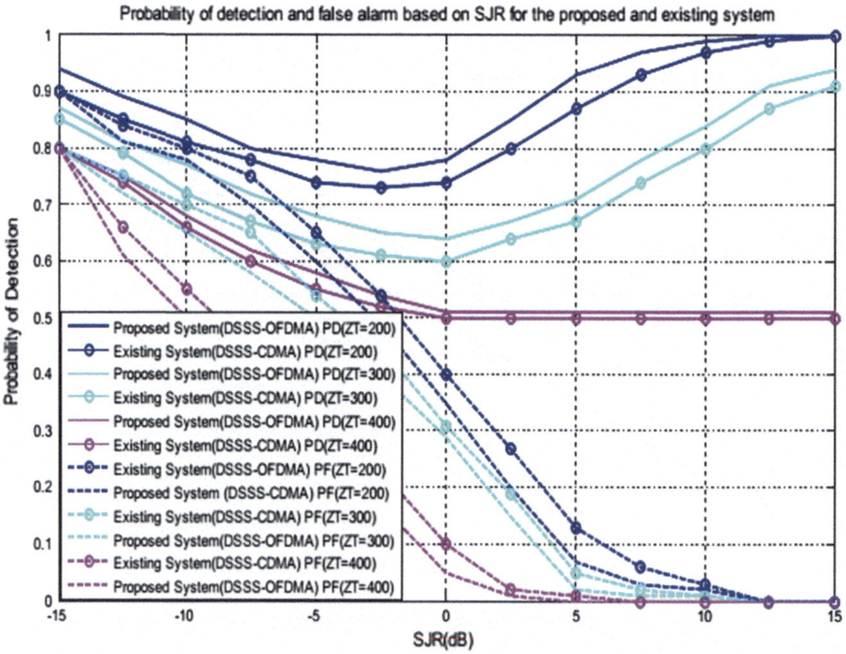

Fig. 5 Signal detection-based SJR for the proposed and existing systems with an interceptional attack

from Fig. 5, the proposed system is outperforming than conventional chaotic system with an interceptional attack. As for high power jammer, the signal detection has been improved and it is reduced while increasing the fixed point under jamming.

6 Conclusion

The system performance using quadratic chaotic sequences has been examined for the projected system with an interceptional attack. For the increasing values of SJR and threshold (Z_T), signal efficiency is improved than the existing system. The aimed DSSS-OFDMA system detecting the signal under jamming is 25% outperforming than conventional chaotic system.

Thus quadratic chaotic sequence is having less complexity with more security and high speed than traditional system. Therefore, from the simulation analysis, it is proven that detecting the signal using quadratic chaos is more effectual than the traditional DSSS-CDMA system.

References

1. Tayebi A, Berber S, Swain A (2014) Performance analysis of chaos based WSN under jamming attack. ISITA, Melbourne
2. Berber S, Feng S (2015) Theoretical modeling and simulation of a chaos-based physical layer for WSN. IEEE
3. La Pan MJ, Clancy TC, Mc Gwier RW (2014) An assessment of OFDM carrier frequency offset synchronization security for 4G systems. In: Military communications conference (MILCOM). IEEE, pp 473–478
4. Kim S, Bok J, Ryu H (2013) Performance evaluation of DCSK system with chaotic maps. 978-1-4673-5742-5/13/$31.00. IEEE
5. Billa R (2012) Analysis of direct sequence spread spectrum over flat fading channel for multiple access. IJCEM, vol 15, issue 5

Mean Availability Parameter-Based DDoS Detection Mechanism for Cloud Computing Environments

Arjunan Amuthan and Pillutla Harikrishna

Abstract Trustworthiness of edge routers and clients plays a significant role in a cloud environment for ensuring reliable packet delivery. Trust of clients depends on the level of cooperation attributed by them for ensuring seamless service and on the support rendered by them for the sake of their neighbouring clients towards the core objective of reliable data dissemination. The level of collaboration between clients is highly influenced by distributed denial of service (DDoS) attacks as they directly influence the performance of cloud computing environment by preventing them from involving in normal data transactions that could result in reduced throughput and packet delivery rate. A mean availability parameter-based DDoS detection mechanism (MAPDDM) is contributed for handling the impacts induced by DDoS towards the dynamic clients of the subnet. The performance of MAPDDM is analysed by varying the size of subnets and number of attackers under the dynamic influence of varying traffic request using CloudSim. The simulation results infer that MAPDDM is phenomenal in sustaining the trust value of clients to a maximum of 82% even when the amount of traffic is varied.

1 Introduction

Cloud computing is considered as the most predominant networking technologies that aims in ensuring seamless cost-effective service. They are capable of combining the merits of other ascendant technologies available in the recent world [1]. They are also potent in facilitating the aspect of multi-tenancy that can share each available resource to multiple connected clients under the impact of maximum traffic load for achieving optimal resource sharing with effective cost [2]. But, the dimension of multi-tenancy imposes several security problems like data availability

A. Amuthan · P. Harikrishna (✉)
Department of CSE, Pondicherry Engineering College, Puducherry, India
e-mail: pillutlaharikrishna@pec.edu

A. Amuthan
e-mail: amuthan@pec.edu

A. M. Zungeru et al. (eds.), *Wireless Communication Networks and Internet of Things*, Lecture Notes in Electrical Engineering 493,
https://doi.org/10.1007/978-981-10-8663-2_12

that prevents the access of data stored at the service provider's end. A number of security threats like the man in the middle, ARP poisoning, ping of death attack and DDoS influence data availability [3]. From the literature, DDoS is identified as one of the potential threats that get easily launched into the subnet and one that is difficult to be prevented [4]. In this paper, MAPDDM is mainly proposed for preventing DDoS attacks by quantifying the mean availability parameter that determines the level of support contributed by the participating edge routers and clients. MAPDDM is also proposed for investigating the influence of varying DDoS attackers on the subnet and also analyses their impact based on varying level of traffic request. The remaining sections of the paper are organized as follows. Section 2 presents a short view on some of the related work that is contributed for handling DDoS available in the literature. The step-by-step process of MAPDDM mechanism in quantifying mean availability parameter for DDoS detection is detailed in Sect. 3. Section 4 highlights the inference of results and discussions that portray the performance of the proposed MAPDDM. Section 5 depicts the major contribution of MAPDDM approach.

2 Related Work

From the recent decade, a number of DDoS detection approaches have been proposed for cloud computing environments [5]. Some of the significant approaches are highlighted with their merits and demerits.

Initially, Janczewski [6] inferred that by preventing DDoS attack, flood based attacks can be filtered out and also help in processing legitimate traffic. Authors implemented a buffer-filter component that stores and filters traffic depending on the bandwidth size of the incoming packets. The buffer-filter component is only activated when the bandwidth size of incoming packets exceeds the threshold level as the additional size of bandwidth is reanalysed for confirming DDoS attacks in the cloud subnet. Chen and Hwang [7] contributed an anti-DoS (AID) technique that possesses the capability of creating overlay network for handling the incoming legitimate data packets. AID Technique also uses special filtering process for processing attack traffic that protects the server from future access.

Further, Du and Nakao [8] proposed a credit-based system that rewards the legitimate behaving users with maximum credit points and punishes the maliciously behaving users by decrementing the credit points they possess. The path isolation through a secure channel is accommodated in this mechanism when cooperative users possess a credit points that significantly exceed the optimal threshold credit points. In this technique, the users are blocked and prevented from server access when the user-contributed credit point is exhausted. This DDoS prevention scheme also employs some of the vital detection components like signalling mechanism and one-hop path splicer and credit counting system. Varalakshmi and Selvi [9] contributed a DDoS defence mechanism that incorporates hop-count filter, normal profile creator and attacker profile creator that aids in comparing data normal traffic to minimize the false positive and false negative rate for facilitating an improved efficiency in attacker detection process through the method of Kullback–Leibler divergence. This DDoS defence mechanism collects, distributes and audits the incoming network traffic for malicious activity detection. This detection methodology detects malicious activity and informs the other clients of the network through reports and alert messages.

Nesmachnow et al. [10] propounded a solution for synchronizing independent tasks of data computing in heterogeneous cloud environments. This synchronizing methodology relies on a licence issue-based management that confirms the authentication of privileged users for remote server access. Authors confirmed that this synchronization-based DDoS handling technique is equally potential with the existing attack detection techniques proposed in the literature. Raj Kumar and Selvakumar [11] proposed a DDoS detection scheme that employs neural classifier for collecting information that highlights and investigates the feature of incoming and test sample traffic. If a deviation between the incoming and sample traffic is detected, the impacts of malicious behaviour are categorized into true positive and true negative based on framed alternatives of false positive and false negative for enhancing the classification accuracy for detection.

3 Proposed Work

MAPDDM is mainly proposed for preventing DDoS attacks by estimating the level of participation in enabling seamless services through the estimation of mean availability parameter that significantly quantifies the degree of support contributed by the participating edge routers and clients. The steps involved in the computing of mean availability parameter are discussed below. In a cloud computing environment, let S_a = {FTN, HTN, STN, LTN, DTN} represent the set of Fully Trusted, Highly Trusted, Semi-Trusted, Low Trusted and Distrusted clients of the considered subnet controlled by a number of edge routers. The trustworthiness of clients is updated to the edge router with the support of comparator. Let $S_b = \{1, 0.75, 0.5, 0.25, 0\}$ denote the set of values returned by neighbours gathered through the aid of probe packets. The probe packets aid in elucidating the availability of the clients as monitored by their neighbours. Availability of clients is explored since DDoS attacks on clients in the cloud environment minimize the degree of services rendered by them to their associated and connected neighbouring clients. The availability of clients is gauged based on 'T_i' and 'D_i' that represents the functioning time and downtime of clients for each client 'i.' Thus, the availability of clients in a cloud environment can be represented through the sequence of independent random variables

$$\{A_i = T_i + D_i\}, \text{ where } i = 1, 2, \ldots .n$$

In this context, $T_1, T_2, \ldots, T_n$ are identically distributed with a general cumulative distributed function $P(t)$ and probability density function $p(t)$. Likewise,

D_1, D_2,..., D_n are also identically distributed with a cumulative density function $F(t)$ and probability density function $f(t)$. Then, A_i which depends on 'T_i' and 'D_i' are also identically and independently distributed. Hence, $\{A_i/i = 1, 2,..., n\}$ is defined as a renewal process. This process is an alternating renewal process as it models five states of the clients. The availability of client (A_i) at any time 't' is quantified based on the sum of expected T_i and expected D_i.

$$A_i = E(A_i) \tag{1}$$

As X_i depends on 'T_i' and 'D_i'

$$A_i = E(T_i + D_i) \tag{2}$$

This availability ($A_i(s)$) corresponds to the convolution of $P(t)$ and $F(t)$ distributions (as $P(t)$ and $F(t)$ are independent)

$$A_i(s) = L_p(s) + L_f(s) \tag{3}$$

where

$$L_p(s) = \int_0^\infty \mathrm{e}^{-sx} p(x)\mathrm{d}x \quad \text{and}$$

$$L_f(s) = \int_0^\infty \mathrm{e}^{-sx} f(x)\mathrm{d}x$$

Then, the convolution property of transform $L_p(s)$ and $L_f(s)$ in terms of relation using

$$L_f(s) = L_p(s) + L_f(s).L_p(s) \tag{4}$$

$$L_f(s) = \frac{L_p(s)}{1 - L_p(s)} \tag{5}$$

Similarly,

$$L_p(s) = \frac{L_f(s)}{1 - L_f(s)} \tag{6}$$

The mean availability of client ($L_m(s)$) in a cloud environment depends on

$$L_m(s) = \frac{L_f(s).L_p(s)}{1 - L_f(s).L_p(s)} \tag{7}$$

where $0 \leq L_m(s) \leq 1$.

Based on $L_m(s)$, the availability of clients is graded as

$$\mathrm{FTN} = (L_m(s), 0.90 < L_m(s) \leq 1.0)$$

$$\mathrm{HTN} = (L_m(s), 0.75 < L_m(s) \leq 0.90)$$

$$\mathrm{STN} = (L_m(s), 0.25 < L_m(s) \leq 0.75)$$

$$\mathrm{LTN} = (L_m(s), 0 < L_m(s) \leq 0.25)$$

$$\mathrm{DTN} = (L_m(s) = 0)$$

Based on the trust evaluated, the clients are identified as DDoS compromised and proper isolation process is triggered.

4 Simulation Experiments and Discussions

The performance of MAPDDM is evaluated using CloudSim simulator by launching attacks through DARPA data set. This experimental analysis is carried out with two, three and four attackers in the subnet of a cloud computing environment. The DARPA data set contains both training set and test data set of size ranging from 300 to 700 bytes in length. The results are validated using both training set and test data set that confirms the evaluation of attack behaviour. In this analysis, traffic generator tool (TG) is used for generating data packets of length 200–6,000 bytes. Three experiments were conducted with two attackers, three attackers and four attackers in the cloud topology. The validation of MAPDDM is ten-fold cross tested under the aggregate view of ten simulation runs of 150 s time period.

A. **Experiment 1: System with Two Attackers**

In experiment 1, each client in the system is made to generate data at a dynamic rate that lies between 150 and 200 packets under an interval of 750–1250 ms. The interval between successive request packets is about 1200 bits, and two client nodes are considered to generate the attack traffic at the maximum rate of 125 request packets each of 120 bits with 2 ms interval. The variations in trust are recorded for a period of 90 s with mean flow rate of 2 m/s at the router with an interval time period of 10 s. In this investigation, the deviation in trust components that emerges to the detection of two attack clients is depicted. In this case, the trusted client node 2 with data rate of 25 Mbits/s is found to possess an average trust value of 0.82. Similarly, the attacker client node 4 with data rate 200 Mbits/s and node 5 with 250 Mbits/s are found to possess an average trust value of 0.26 and 0.28, respectively. It is proved that MAPDDM is significant than the AID technique used for comparative analysis as it has the capability of sustaining DDoS attacks to a maximum of 26%.

B. **Experiment 2: System with Three Attackers**

In experiment 2, each client in the system is made to generate data at a dynamic rate that lies between 200 and 400 packets under an interval of 750–1250 ms. The interval between successive request packets is about 1200 bits, and three client nodes are considered to generate the attack traffic at the maximum rate of 150 request packets each of 150 bits with 2 ms interval. The variations in trust are recorded for a period of 90 s with mean flow rate of 2 m/s at the router with an interval time period of 10 s. In this analysis, the deviation in trust components that emerges to the detection of three attack clients is portrayed. In this case, the trusted client node 2 with data rate of 25 Mbits/s is found to possess an average trust value of 0.73. Similarly, the attacker client node 4 with data rate 200 Mbits/s, node 5 with 250 Mbits/s, and node 7 with data rate 400 Mbits/s are found to possess an average trust value of 0.22 and 0.19 and 0.21, respectively. It is proved that MAPDDM is significant than the AID technique used for comparative analysis as it has the capability of sustaining DDoS attacks to a maximum of 22%.

C. **Experiment 3: System with Four Attackers**

In experiment 3, each client in the system is made to generate data at a dynamic rate that lies between 300 and 500 packets under an interval of 1000–1450 ms. The interval between successive request packets is about 1350 bits, and four client nodes are considered to generate the attack traffic at the maximum rate of 200 request packets each of 200 bits with 2 ms interval. The variations in trust are recorded for a period of 120 s with mean flow rate of 2 m/s at the router with an interval time period of 10 s. The deviation in trust components that emerges to the detection of four attack clients is explained in this exploration. In this case, the trusted client node 2 with data rate of 25 Mbits/s is found to possess an average trust value of 0.71. Similarly, the attacker client node 4 with data rate 200 Mbits/s, node 5 with 250 Mbits/s, node 7 with data rate 400 Mbits/s, and node 8 with data rate 400 Mbits/s are found to possess an average trust value of 0.22, 0.16, 0.19 and 0.17 respectively. It is proved that MAPDDM is significant than the AID technique used for comparative analysis as it has the capability of sustaining DDoS attacks to a maximum of 19%. It is also proved that MAPDDM is significant than the existing DDoS handling techniques as it is capable of improving the throughput and packet delivery rate of cloud computing environment. Finally, Figs. 1 and 2 highlight the performance of MAPDDM based on throughput and packet delivery ratio.

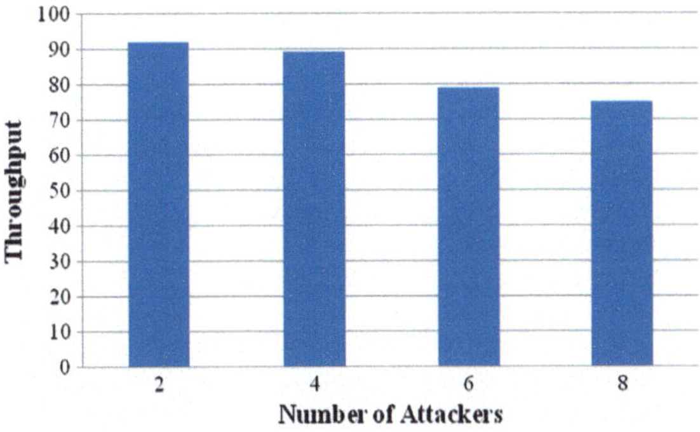

Fig. 1 Performance of MAPDDM based on throughput

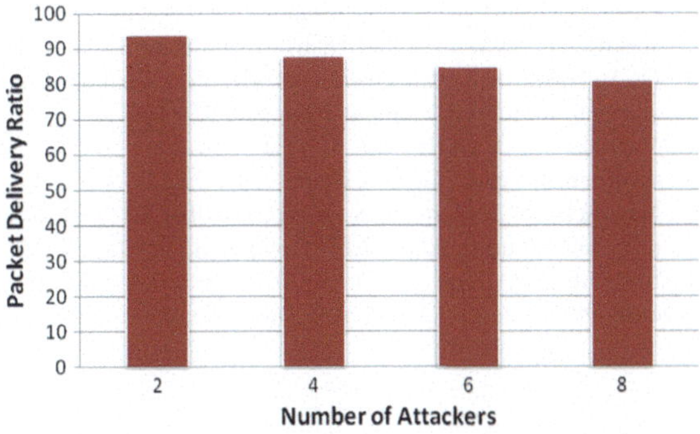

Fig. 2 Performance of MAPDDM based on packet delivery ratio

5 Conclusion

In this paper, MAPDDM is proposed for detecting malicious behaving edge routers and client nodes by incorporating a rapid DDoS detection process since large amount of huge requests that enter as attack traffic has to probe before it enters the network. The main contribution of the MAPDDM lies in its potentiality of classifying clients as Fully Trusted, Highly Trusted, Semi-Trusted, Low Trusted and Distrusted client by computing mean availability parameter using functioning and downtime. The simulation results confirm that MAPDDM is highly survivable to DDoS attacks as the measured false positive and false negative are negligible. This negligible measure is due to the facilitation of lower sensitive detection time for accurately confirming an edge router or client node as a trusted entity or an attacker.

References

1. Kim SS, Reddy ALN (2008) Statistical techniques for detecting traffic anomalies through packet header data. IEEE/ACM Trans Netw 16:562–575
2. Lopez R, Onate E (2006) A variational formulation for the multilayer perceptron. In: Proceeding of artificial neural networks—ICANN 2006, Lecture Notes in Computer Science, Athens, pp 159–168
3. Modi CN, Patel DR, Patel A, Rajarajan M (2012) Integrating signature Apriori based network intrusion detection system (NIDS) in cloud computing. Procedia Technol 6:905–912
4. Lo C-H, Ansari N (2013) CONSUMER: A novel hybrid intrusion detection system for distribution networks in smart grid. IEEE Trans Emerg Topics Comput 1:33–44
5. Li M, Li M (2010) An adaptive approach for defending against DDoS attacks. Math Probl Eng 66:1137–1151
6. Janczewski LJ (2001) Handling distributed denial-of-service attacks. Inf Secur Tech Rep 6:37–44
7. Chen Y, Hwang K (2006) Collaborative detection and filtering of shrew DDoS attacks using spectral analysis. J Parallel Distrib Comput 66:1137–1151
8. Du P, Nakao A (2010) OverCourt: DDoS mitigation through credit-based traffic segregation and path migration. Comput Commun 33:2164–2175
9. Varalakshmi P, Selvi ST (2013) Thwarting DDoS attacks in grid using information divergence. Future Gener Comput Syst 29:429–441
10. Nesmachnow S, Iturriaga S, Dorronsoro B (2015) Efficient heuristics for profit optimization of virtual cloud brokers. IEEE Comput Intell Mag 10:33–43
11. Raj Kumar PA, Selvakumar S (2011) Distributed denial of service attack detection using an ensemble of neural classifier. Comput Commun 34:1328–1341

An Effective Dynamic Slot Allocation Scheme for Wireless Body Area Network

M. Ambigavathi and D. Sridharan

Abstract Wireless body area network (WBAN) is an indispensable research field in recent times, in order to monitor and transfer the lifesaving or critical data over wireless medium. One crucial challenge for WBAN is to track and sustain the Quality of Service (QoS). Another significant issue is to reduce the energy consumption within such a resource-constrained network. Due to consecutive time slots allocation, the ready nodes in the queue have to wait for a long time even it has very few or bursty data for transmission. The fixed slot assignment is not suitable for the emergency applications where nodes have different packet generation rate. Therefore, this paper introduces a novel enhanced time division multiple access (ETDMA) MAC protocol using IEEE 802.15.4 standard which dynamically allocated the time slots for body sensor nodes. The entire channels are divided into a number of time slots, in which nodes can send their control or data packets over delegated time slot. Also, the Enhanced Packet Scheduling Algorithm (EPSA) is considered to ensure the great number of lingering or waiting nodes allocated to the specified time slots in the given time frame. Since, it avoids unnecessary bandwidth usage and channel assignment. The simulation results show that the proposed method saves more energy, reduce the delay and packet loss, and employs the channel to maximum capacity.

1 Introduction

Wireless body area network is a human-centric network recently which creates special attention in the field of health monitoring applications. This network consists of low-power, smart sensor nodes dispersed on, in or around the human body for observing the vital parameters [1]. One of the greatest challenges is to reduce the energy consumption. Particularly for the nodes which are implanted into the human body because that battery replacement or recharging is impractical. Therefore, the

M. Ambigavathi (✉) · D. Sridharan
Department of ECE, Anna University, CEG Campus, Chennai, India
e-mail: ambigaindhu8@gmail.com

A. M. Zungeru et al. (eds.), *Wireless Communication Networks and Internet of Things*, Lecture Notes in Electrical Engineering 493,
https://doi.org/10.1007/978-981-10-8663-2_13

design of energy efficient MAC protocol is the major concern to minimize the energy wastage of body sensor nodes. IEEE 802.15.4 [2] is a low-power communication standard that describes the specifications for Physical layer (PHY) and MAC layer of a low data rate WBAN with low-latency requirements. Random Access MAC protocols (CSMA/CA) and Schedule-based MAC protocols (TDMA) are the two major medium access schemes in WBAN. Sensor nodes compete for channel access prior to data transmission in CSMA/CA. In TDMA method, node can only transmit its data in an assigned time slot. In this type, many nodes can get the chance to transmit their sensed data simultaneously without any collisions [3]. In which, time base is usually divided into the equal number of time slots, after these slots are further organized into number of superframes. Within each frame, a node is occupied at least one time slot to transmit its data. Hence, it has a natural advantage of collision-free medium access [4].

The major concern in TDMA-based scheduling is synchronization of the nodes with a network coordinator when node detects abnormal value and tries to schedule the time slots access the medium topological changes. Since, there is necessary to be considered these possibilities when designing a new MAC protocol. The proposed technique competently works without any earmarked allocation of channels in order to avoid wastage of the time slots and shows the improved performance results with varied data types.

The rest of the paper is outlined as follows: Represents the related works and the proposed method is explained in Sect. 3. The simulation results are evaluated in Sect. 4. Finally, this paper is concluded in Sect. 5.

2 Related Work

This section recaps the current issues of TDMA-based MAC protocols and methods with neat focus on energy consumption, packet collision, and delay. Authors in [5] proposed a novel scheme to reduce unnecessary usage of bandwidth by allocating the Guaranteed Time Slot (GTS) to body sensor nodes with respect to GTS duration until the transmission of sensed data to the coordinator node. It saves sufficient bandwidth in the Contention Free Period (CFP), and then it is assigned to the Contention Access Period (CAP), thus increased the whole network's performance slightly. In [6], a Binary Integer Nonlinear Dynamic Programming (BINDP) slot allocation scheme has introduced to improve the energy efficiency. This technique also jointly considered two heuristic algorithms to assign the data priorities in order to maximize the slot utilization with less delay and resolved the queue limitations. It improves the energy efficiency, delay, and buffer limitations are also fulfilled. Figure 1 represents the overall architecture diagram of WBAN system. iQueue-MAC has developed to deal with burst or heavy traffic. It has jointly considered both CSMA/CA and TDMA types of channel access MAC protocols for packet transmission. Therefore, dynamically schedules the time slots based on the different traffic rate. This method used a queue indicator that is piggybacked onto all

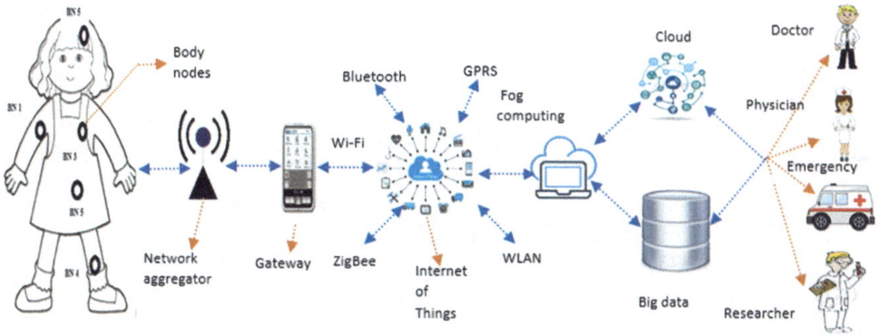

Fig. 1 Overall architecture diagram of WBAN

data packets, with time slot requests for the channel allocation to the nodes. iQueue-MAC operates the bandwidth more precisely, allocates the required slots, and retransmits the packets using TDMA mechanism for intensive body sensor nodes [7]. Authors presented a novel dynamic channel allocation mechanism to solve the collision problem. It uses flexible allocation field to handle different conditions of the physiological data. If there is any packet collision, then it starts with a reallocation process to permit the failed packets with minimum retransmission delay. Hence, the packet delivery delay and unwanted usage of bandwidth are minimized in [8]. Authors considered the greedy technique with Improved Packet Scheduling Mechanism (IPSM) for dynamic allocation of time slots for waiting nodes in the given frame and also improved the overall performance of the network with efficient channel utilization [9].

3 Proposed Method

Time Division Multiple Access (TDMA) is a classical channel access method for WBAN. The main drawbacks of static TDMA are the lower channel utilization rate in the presence of heterogeneous traffic loads, not capable of reducing the packet loss rate and delay in case of packet collisions. To minimize such issues, this paper introduces a novel Enhanced Packet Scheduling Algorithm (EPSA) for scheduling the data packets in order to minimize the energy consumption and increase the overall throughput. The main condition is that each node must know the information of all other neighbors. This EPSA performs scheduling process for the nodes which senses data and ready to have a certain time frame. If node sensed the data, it has received a certain time slot in a given frame, sometimes it may be possible that some nodes have data to transmit but still not get a time slot in that frame. In another case, some time slots are sent empty. To avoid such issues and improve channel utilization, this algorithm schedules time slots before transmitting the frame and make available empty slots to other waiting nodes or ready nodes

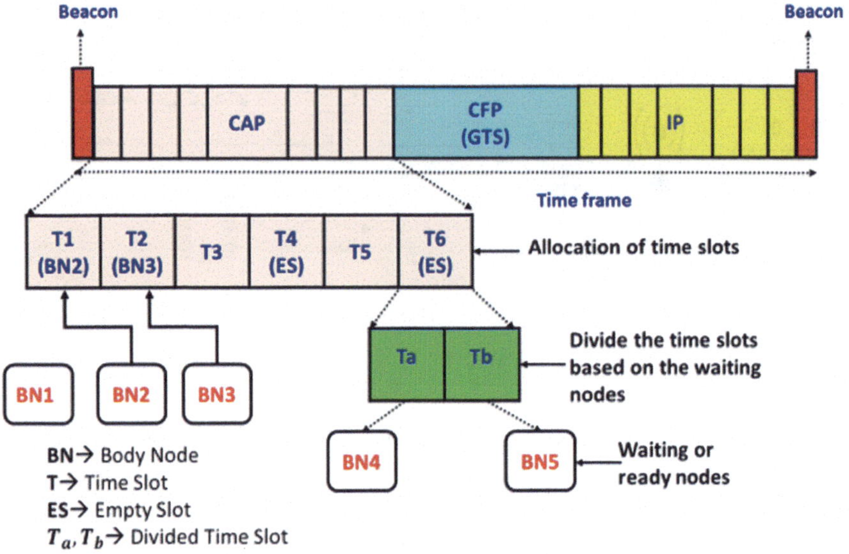

Fig. 2 Allocation and division of time slots

with vital information. Moreover, the body sensors are not required to wait for the next superframe time slot. Figure 2 represents the pictorial representation of proposed method.

3.1 *Allocation of Time Slots for Body Nodes*

In this step, Body Node (BN1) has been assigned a time slot T1. At the same time, T1 is occupied by BN2 and similarly time slot T2 by BN3. So, then BN1 can be assigned an empty slot either T4 or T6, and also nodes are not required to wait for next time frame in order to transmit the vital information which the node has sensed. Sometimes, many nodes have vital data that are waiting in queue for the time slots means, then node observes which slots are empty in a certain time frame so that it can be assigned to the ready nodes or waiting nodes.

By using this identification, the empty slots are sent in a given frame has reduced significantly, even if there are some waiting nodes or ready nodes available for sending the vital information.

3.2 Division of Empty Slots for Ready Nodes

This second step is carried out when there are number of waiting nodes or ready nodes that want to transmit the vital data, but the time slots are not restricted. At this scenario, the remaining unused or empty time slots are divided into smaller equal length of slots based on the number of waiting nodes. Thus, the number of body nodes can send their vital data at once without any time delay and also it saves more energy. Here, the T4 is the empty slot and then divide this slot into two equal time slots and now it has two slots in the given time frame. Hence, two nodes can able to send the vital data instead of one node. Algorithm 1 explains the entire process of proposed technique.

Algorithm 1 Enhanced Packet Scheduling (EPSA)

Step 1: Initialize the parameters for number of waiting sensor nodes (N_s), empty time slots (T_e), time slots (T_s), waiting for next frame (W_f)
Step 2: Identify the T_s
Step 3: Determine the empty time slots (T_e) available in a given time frame
Step 4: Check the condition If T_s is T_e then
Step 5: Assign next time frame for waiting nodes (W_f)
Step 6: Else set W_f
Step 7: End if
Step 8: End else
Step 9: If $T_e \geq W_f$ then
Step 10: Assign T_e
Step 11: Else divide T_e
Step 12: Set FIFO End if

4 Results and Discussion

This section describes the simulation results of proposed technique. The novel idea of this work is also compared with existing method by considering the following parameters such as packet delivery ratio, delay, throughput, and power consumption.

Table 1 illustrates the basic parameter's details for simulation.

4.1 Packet Delivery Ratio (PDR)

Figure 3 shows the simulation analysis of packet delivery ratio of proposed technique. The packet delivery ratio is calculated between the number of packets transmitted by the body sensor nodes and the number of packets received by a coordinator node. In this graph, the packet delivery rate is higher than IPSM technique.

Table 1 Simulation parameters

Parameters	Values
Deployment area	400 × 400 m
Queue limit	50
Application type	Event driven
Packet size	32 bytes
MAC	IEEE 802.15.4
Number of nodes	5
Simulation tool	NS2
Vital parameters	Temperature, SPo2, BP, HR, and RR

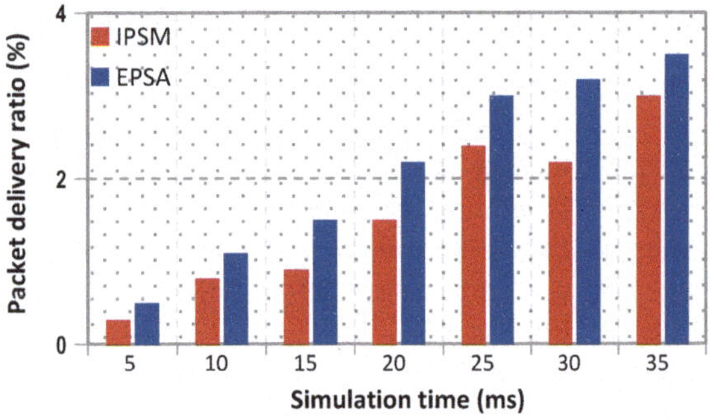

Fig. 3 Packet delivery ratio

4.2 Delay

The average time delay is the time taken by the packets to reach a coordinator node from the body sender nodes. Figure 4 shows the delay of the waiting nodes. Figure 4 depicts the packet size in *x*-axis with respect to simulation time and the delay in *y*-axis (milliseconds). From this evaluation, it is analyzed that the delay of proposed technique is moderate when compared to the existing method.

4.3 Power Consumption

Figure 5 demonstrates the power consumption of both existing and proposed methods. Figure 5 indicates the simulation time (ms) in *x*-axis and the power consumption (mW) of the nodes in *y*-axis. The proposed technique has minimum power consumption when compared with existing technique by the avoiding unnecessary slot allocations.

Fig. 4 Delay

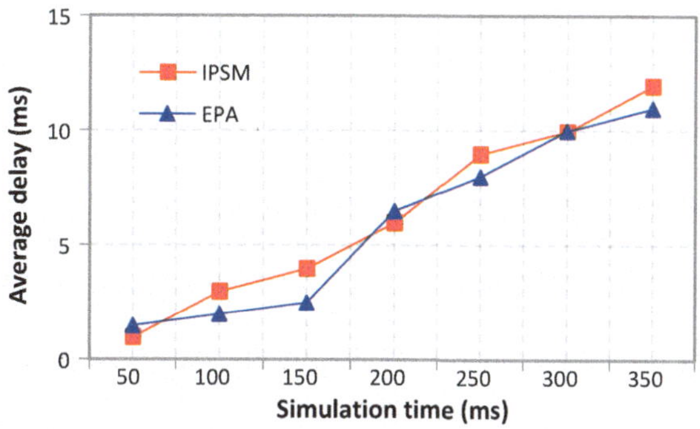

Fig. 5 Power consumption

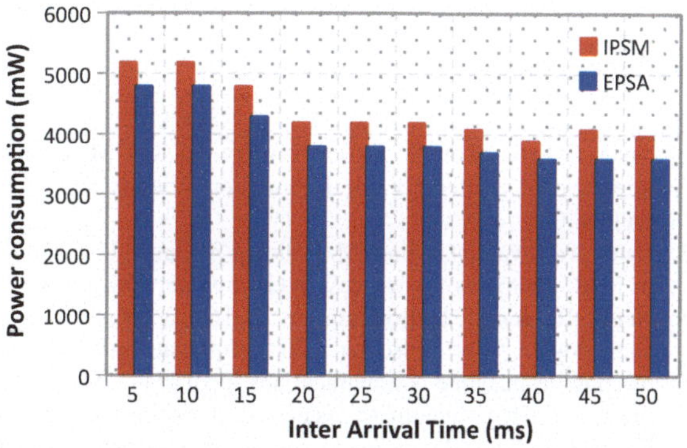

4.4 Throughput

Figure 6 shows the throughput of simulation analysis. The overall throughput is achieved by calculating the total number of bits can be received from body sensor nodes to a coordinator node in a given amount of time. Here, the overall throughput is high compared to existing scenario.

Fig. 6 Throughput

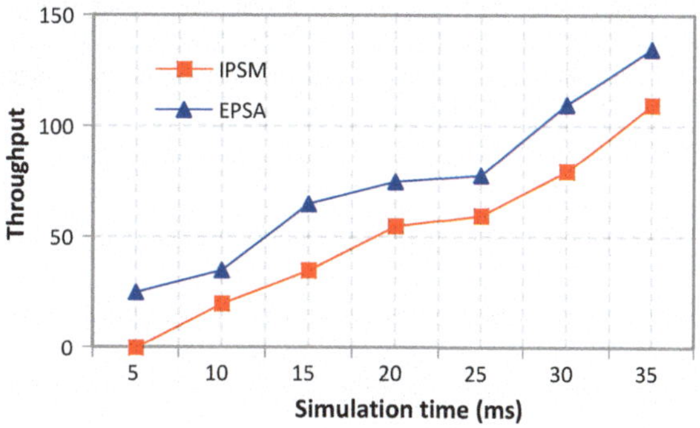

5 Conclusion

This paper proposed a novel dynamic Enhanced TDMA (ETDMA) MAC protocol using IEEE 802.15.4 standard. Also, the Enhanced Packet Scheduling Algorithm (EPSA) has introduced for dynamic allocation of time slots to the body sensor nodes. By using this method, the time slots are dynamically allocated to the waiting nodes or ready nodes in an available time frame that significantly reduced the power consumption and also improved the overall performance by utilizing the channels to its maximum capacity. Consequently, delay and packet loss can be greatly reduced. The future work includes implementing the superframe structure with priority assignment for waiting nodes by modifying this mechanism.

References

1. Yuan J, Li C, Zhu W (2013) Energy-efficient MAC in wireless body area networks. In: International conference on information science and technology application (ICISTA-13), pp 021–024
2. Rahim A, Javaid N, Aslam M, Rahman Z, Qasim U, Khan ZA (2012) A comprehensive survey of MAC protocols for wireless body area networks. arXiv:208.2351v1
3. Negra R, Jemili I, Belghith A (2016) Wireless body area networks: applications and technologies. In: Elsevier 2nd international workshop on recent advances on machine-to-machine communications, pp 1274–1281
4. Sruthi R (2016) Medium access control protocols for wireless body area networks: a survey. In: Elsevier global colloquium in recent advancement and effectual researches in engineering, science and technology (RAEREST 2016), pp 621–628
5. Goyal R, Patel RB, Bhadauria HS, Prasad D (2014) Dynamic slot allocation scheme for efficient bandwidth utilization in wireless body area network. In: 9th IEEE international conference on industrial and information systems (ICIIS). ISSN: 2164-7011
6. Kong R, Chen C, Yu W, Yang B, Guan X (2013) Data priority based slot allocation for wireless body area networks. In: International conference on wireless communications and signal processing (WCSP). https://doi.org/10.1109/wcsp.2013.6677273
7. Zhuo S, Wang Z, Song YQ, Wang Z (2013) iQueue-MAC: a traffic adaptive duty-cycled MAC protocol with dynamic slot allocation. In: 10th annual IEEE communications society conference on sensor, mesh and ad hoc communications and networks (SECON), pp 95–103
8. Pourmohseni B, Eshghi M (2013) Reliable energy efficient dynamic-TDMA MAC protocol for wireless body area networks. Int J Appl Innov Eng Manage (IJAIEM) 2(7):393–402
9. Unk N, Trivedi A, Razaque A (2016) Dynamic allocation of slot using MAC protocol. In: IEEE conference long island systems, applications and technology (LISAT). https://doi.org/10.1109/lisat.2016.7494147

Enhancement of Security and Confidentiality for D2D Communication in LTE-Advanced Network Using Optimised Protocol

Payal P. Tayade and Perumal Vijayakumar

Abstract Third Generation Partnership Projects (3GPPs) introduced fourth generation of wireless communication entitled Long-Term Evolution-Advanced (LTE-A). Security always plays a vital role in communication. Device-to-device (D2D) communication in LTE-A in between user equipments can be carried out using Secure Data Sharing Strategy (SeDS) via evolved node B (eNB) and Gateway (GW). But, communication between eNB and GW is not secure. And if this part of system will not be secure, further communication may get hampered. In this paper, implementation of optimised protocol has been shown. The proposed protocol enhances security in data communication between the nodes.

Keywords Long-term evolution-advanced (LTE-A) network · Security
Device-to-device (D2D) · Secure data sharing strategy (SeDS)

1 Introduction

Huge demand of broadband wireless communication data and various multimedia applications obligate to carry out fast improvement in the sector of wireless network. Enormous users want to access the network from any place and at any time via their devices [1]. LTE-A has been developed to fulfil the demand of wireless communication network and its users. But, with the establishment of communication using LTE-A network, security and confidentiality are the key factors which should get focussed. In the year of 2014, a group-based security protocol was proposed for LTE-A but it was for machine-type communication (MTC). The protocol was implemented in concern with security and performance for MTC [2].

P. P. Tayade · P. Vijayakumar (✉)
School of Electronics Engineering, VIT University, Chennai Campus,
Chennai 600127, Tamil Nadu, India
e-mail: Vijaya.kumar@vit.ac.in

P. P. Tayade
e-mail: payalpurushottam.tayade2015@vit.ac.in

A. M. Zungeru et al. (eds.), *Wireless Communication Networks and Internet of Things*, Lecture Notes in Electrical Engineering 493,
https://doi.org/10.1007/978-981-10-8663-2_14

Device-to-device (D2D) communication in cellular network enables to establish direct communication between two mobile users or nearby mobiles without extending across the base station (BS) or core network [3]. Hence, D2D communication is considered as one of the best way out for data offloading. Furthermore, it also delivers some unique features such as making use of information between critical public safety networks and commercial networks based on LTE. Ultimately, it helps to achieve significant performance and efficiency benefits in LTE network. As LTE-A network operates on a licensed band; it delivers a systematic and planned deployment which results in a better user satisfaction and quality of service. Hence, it is keenly required to develop an optimised protocol for the combination of LTE-A network with D2D communication to provide better security [4].

The remainder of the paper is constructed as below. Detailed information about conventional protocol is explained in Sect. 2. The proposed protocol is summarised in Sect. 3. Section 4 contains the details of proposed security protocol. Discussion and simulation result is elaborated in Sect. 5. Finally, Sect. 6 concludes the paper.

2 Previous Research Review

Detailed study on integration of D2D communication in LTE-A network has been done. It has been found out that most of them have researched on service quality, network congestion, pricing scheme, seamless offloading, mobility of relay and joint neighbour parameters. A very few researchers have considered security for research. One associated paper for security in D2D communication with LTA network has revealed that how secure communication can be established in between two user equipments via eNB and GW. But, some loopholes are present in the proposed system which results into lack of confidentiality during the communication.

The most similar kind of study of this paper has been done in various wireless networks such as in wireless body area networks (WBANs) and vehicle ad hoc networks (VANETs). To achieve security in terms of access controls in WBANs, symmetric cryptography is implemented. On the other hand, public key infrastructure (PKI) is responsible to establish security requirements in VANETs. Using combination of both symmetric cryptography and PKI, one protocol is designed to achieve security in D2D communication between two user equipments in LTE-A networks. By considering all the points, we attempt to design one protocol where integration of symmetric cryptography, PKI and elliptic cryptography curve. Analysis of previous papers where D2D communication is implemented in LTE network can be summarised through Table 1.

Table 1 Previous research review of D2D communication with LTE-A network

S. No.	Title of paper	Advantages	Disadvantages
1	When D2D communication improves group-oriented services in beyond 4G networks (2015) [3]	Improving service quality in terms of delay and energy consumption	Security
2	Secure and smart media sharing based on a novel mobile device-to-device communication framework with security and procedures (2015) [4]	Reduced network congestion and better pricing scheme	Security and reliability
3	Secure data sharing strategy for D2D communication LTE-Advanced networks (2015) [5]	Secure data and sharing mechanism	Communication between eNB and GW is not secured
4	Efficient load balancing using D2D Comm. and biasing in LTE-Advance Het-Nets (2015) [6]	Seamless offloading and mobility of the relay	Implemented for one macro base station only
5	Enabling D2D communications through neighbour discovery in LTE cellular networks (2015) [7]	Neighbour discovery and joint neighbour detection	Orthogonality of Ψ cannot be preserved

3 Conventional Secure Data Sharing Protocol System

In 2016, for D2D communication in LTE-A networks, one secure data sharing protocol was implemented. It is also entitled as "Secure Data Sharing Strategy (SeDS)". This system was implemented with the combination of public key cryptography and symmetric encryption. The proposed protocol helps to achieve security and availability parameter in D2D communication. Detail analyses on D2D communication in comparison with WBAN's and VANET's have also been done. For this investigation, two parameters have been focussed entitled as public key infrastructure (PKI) and symmetric encryption. The inference of the analysis is yet in D2D, PKI and symmetric encryption is not implemented. But this is not the case in regard with WBANs and VANETs. Implementation of PKI and symmetric encryption is available in both the cases [8–11].

3.1 Network Architecture and Threats for Conventional Protocol System

Network architecture for SeDS protocol system includes four important parts: gateway (GW), evolved node B (eNB), user equipments (UEs) and service providers (SPs). Out of which, eNB and GW are assumed to be trustworthy which will not get affected by attacker. D2D communication usually gets attacked by free-riding attack, which ultimately reduces the system availability and privacy preservation in terms of security. The proposed SeDS protocol is based on two preliminaries known as Bilinear Pairing and Diffie–Hellman Key Exchange (DHKE) [5].

3.2 Initialization of SeDS Protocol System

System initialization of SeDS protocol is done through four steps such as system parameter generation, SP registration, UE registration and system setup.

System parameter generation step is used for generation of tuple (q, g, $g1$, G, GT, $\acute{e}$) by using function Gen (K). Also, selection of secure symmetric encryption algorithm Enc_s () and two hash function H_0 and H_1 are done by eNB in this step only. In second step that is in SP registration, registration of real identity (RID_0) is done so that it will be able to provide original data in the system. Then, calculation of PID_0 is done using $\text{PID}_0 = H_0$ (RID_0). PID_0 is known as pseudo-identity for SP. Finally, calculation of private and public key will be done (X_0, $x0$) and it will be sent to SP by eNB through secure channel.

To minimise overhead of communication, SP and UE both calculate their pseudo-identity by their own. SP registration will be done exactly similar to UE registration. Moving towards the last step which is system setup, it plays a very important role in system initialisation. In this step, eNB is used to keep a record of various parameters such as RID, PID, public key, portion index (Pi), share frequency and malicious behaviour amount. Also, to check original data, combined record of Pi and payload (M) is stored by eNB. Furthermore, for the sake of data authority and integrity, computation of signature σ_1 is done by SP and will be attached with Pi and M. The whole process is completed online so that data sharing latency will be minimised.

3.3 Sharing of Data Using SeDS Protocol System

Once the initialisation of the system is finished, one needs to focus on how data sharing is carried out using the SeDS protocol system. To understand data sharing, we need to consider two user equipments known as UE_i and UE_j, one eNB and one GW. The whole process is divided into 8 steps.

The first step is known as “Service Request”. In this step, UE_i (who wants to have ith frame of data) chooses c (where $c \in$ to Z_q^*) and calculates key hint z (where $z = g^c$) so that communication key K_c will be generated. Also, this step calculates HMAC (Internet Standard RFC 2104) of message M using hash function of parallel combination of $K^+ \oplus$ opad, $H(K^+ \oplus$ ipad) and m, where

K^+ is the key padded out to size;
opad = 0011 0110, 0011 0110 and so on;
ipad = 0101 1100, 0101 1100 and so on.

Finally, service request message will be sent to eNB, which includes whole HMAC message with its $PID_{i,}$ z and the expected Pi.

The second step is entitled as "Authentication". After receiving request message, authenticity will be checked by eNB by calculating hash value of message. If it is found to be authenticated, it will be checked that whether it is available in record table or not. If it is available it will be ignored, and if not the message will be sent to GW with its RID_i followed by that the third step will be performed which is called as "Candidate Detection". In this step, detection of valid D2D pair is done for requesting UE by Proximity Service Control Function (PSCF) and finally gateway sends to eNB with its RID_j.

The next step is known as "Pair Selection". This step as its name suggests performs the proper selection of the candidate with which Pi will match. Sending of request message to selected entity and acknowledgement of it will also be done in this step only. The real data transmission process starts with next step called as "Data Transmission". Once request message is received, encryption of message will be done to recover original message and it will be sent with signature $\sigma_{2.}$ The whole message is obtained in the format of PID_i, $PID_{j,}$ Pi, Enc (M) (M'), Time Stamp (Ts), Signature (σ_1) and Signature (σ_2). Finally, this message will be sent from transmitter to eNB so that its shared frequency record will be updated.

The further step is called as "Entity Verification". In this step, after receiving packet extraction of PID_j will be done by UE_i from message. Followed by that, comparison of PID_j and pseudo-identity will be done. If match is obtained, packed will be dropped. If not, checking will be done by verifying the balance of the following equation. If encryption of X_j and hash function of parallel combination of PID_j, Pi, M', Ts and σ_1 is equal to encryption of σ_2, g then it confirms that the data is sent by entity with pseudo-identity PID_j. In this step, for the sake of timestamp Ts, eNB verifies that the message is sent within the allowable time window or not. If it is so, decryption of message will be carried out and feedback will be recorded.

The seventh step is known as "Data Verification". To check data authority, encryption of X_0 and hash function of parallel combination of Pi and M is calculated. Also, encryption of σ_1 and g is calculated. If both the values are found to be equal, then the data will be considered to be authorised. And if not, then it is considered as impersonation attack.

The last step is "Record Refresh". Here, eNB verifies the validity of σ_1, and if it is found to the valid, PID_i column in record table will get refreshed by inserting Pi and also shared frequency of PID_j will be incremented by 1, and if σ_1 is not found to be valid, the malicious behaviour amount record will be updated. In this way, whole SeDS protocol system is implemented to achieve security [5].

While dealing with such scenario, the most crucial part is to select proper device to which we want to send information. We have seen the device detection done by GW and the RID of selected device is sent to eNB. But here, the drawback of the system is that the communication between eNB and GW is not secure. It can be shown through Fig. 1. If this part of the system is not secure the further communication will be no more authentic. Hence, to achieve security between eNB and GW, a new idea has been proposed. It is explained in the next section.

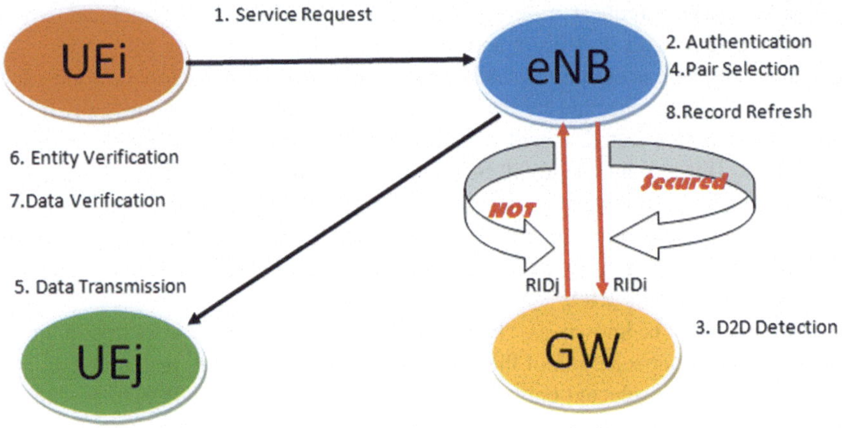

Fig. 1 Conventional SeDS protocol system for D2D communication in LTE-A network

4 Secured Protocol System in the D2D Communication in LTE-A

To achieve security between eNB and GW, we consider the implementation of Elliptic Curve Cryptography (ECC). In the conventional SeDS protocol system, to establish secure communication before D2D detection by gateway can be implemented through following steps.

1. Once service request will be sent from UE_i to eNB, the authentication will be checked by eNB and RID_i will be sent to GW. Also, one random integer $e \in [1, n-1]$ will be selected by eNB. (RID_i, e) which will be sent to GW, where n is the order chosen for ECC operations.
2. Device-to-device detection will be performed by PSCF in GW. Let the real identity generated for the detected device is RID_j. Also, one random integer key $K_G \in [1, n-1]$ will be selected by GW. Finally, $R_G = K_G \chi G$ will be sent to eNB. e and K_G for eNB and GW can be considered as temporary secret key so that it helps to hide their real identity.
3. From the random integer selection, GW will calculate one S_G parameter, where, $S_G = K_G + e \times d_G$, + is used for scalar addition and d_G is the secret key of GW. From II and III step (S_G, RID_j) will be sent to eNB.
4. From received data, eNB will check $R_G = S_G.G - e.Q_G$ where Q_G public key of GW. If obtained R_G and calculated R_G is same, eNB will consider RID_j as the authenticated, and in this way, the secure communication can be established in between eNB and GW [10]. Once RID will be authenticated, further communication between two user equipments can be carried out using the steps of conventional protocol system (refer Fig. 2) [12, 13].

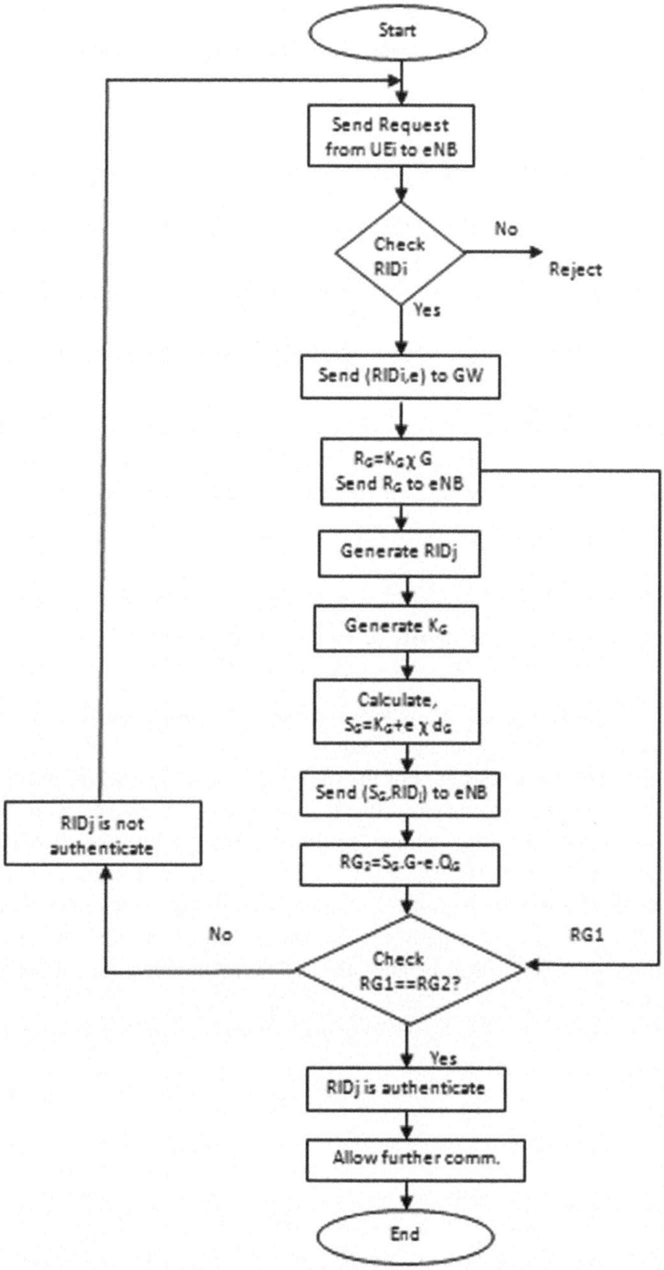

Fig. 2 Flow chart for the proposed authentication protocol

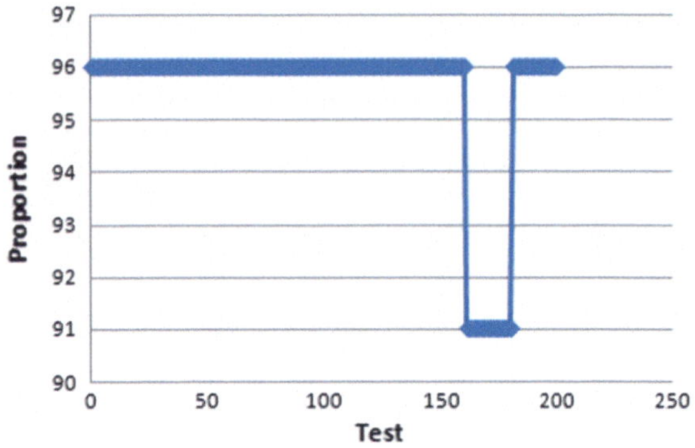

Fig. 3 Plot of proportion versus tests

The methodology that implemented here is the combination of symmetric cryptography, public key infrastructure and ECC. The proposed idea can be implemented on Network Simulator 3 (NS3) software to verify the simulation results.

5 Discussion on Randomness of the Generated Key

When ECC is implemented in between eNB and GW to achieve security, its intensity depends on the generated random key. Hence, it is necessary to check proportion of randomness of key. For the randomness properties of produced key, it needs to get accurate statistical results. If 100 binary sequences were tested, 96 binary sequences had *P*-values which will be less than or equal to 0.1. Hence, the proportion obtained will be 96/100 = 0.96. NIST, a statistical test is used to analyse the randomness of the key. Figure 3 indicates how many samples have passed the given tests. The following simulation results have shown that the generated random key sequences pass all the tests and ultimately maintained its randomness and uniformity [14].

6 Conclusion

We have proposed an additional security system in SeDS protocol system to establish secure communication in between eNB and GW for D2D communication in LTE-A network. The system is explicitly designed to achieve authenticity in the communication while detecting proper device by the GW. The conventional SeDS protocol was implemented using digital signature and symmetric encryption. Furthermore, we have included ECC so that even security between eNB and GW could not get compromised.

We have analysed the D2D communication scenario and proposed above idea but its validity needs to be checked by implementing simulation on NS3 platform. The limitation to this paper is it needs to be checked whether the idea could get implemented when more than two user equipments are there.

References

1. Jengyueng C, Chunchuan Y, Yiting M (2015) A novel smart forwarding scheme in LTE-Advanced networks. China Commun 12(3):120–131
2. Choi D, Choi HK (2013) An group-based security protocol for machine type communications in LTE-Advanced. J Korea Inst Inf Secur Cryptology 23(5):885–896
3. Militano L, Condoluci M, Araniti G, Molinaro A, Iera A (2014) When D2D communication improves group oriented services in beyond 4G networks. Wireless Netw 21(4):1363–1377
4. Ramasubramanian S, Chung S, Ryu S, Ding L (2015) Secure and smart media sharing based on a novel mobile device-to-device communication framework with security and procedures. ACM 35–49
5. Zhang A, Chen J, Hu R, Qian Y (2016) SeDS: secure data sharing strategy for D2D communication in LTE-Advanced networks. IEEE Trans Veh Technol 65(4):2659–2672
6. Sharma A, Trivedi A, Roberts N (2015) Efficient load balancing using D2D communication and biasing in LTE-Advance Het-Nets. ACM
7. Tang H, Ding Z, Levy B (2014) Enabling D2D communications through neighbor discovery in LTE cellular networks. IEEE Trans Signal Process 62(19):5157–5170
8. Lu R, Lin X, Shen X (2013) SPOC: a secure and privacy-preserving opportunistic computing framework for mobile-healthcare emergency. IEzaEE Trans Parallel Distrib Syst 24(3):614–624
9. Mana M, Feham M, Bensaber B (2011) Trust key management scheme for wireless body area networks. Int J Netw Secur 12(2):75–83
10. Liang X, Li X, Lu R, Lin X, Shen X (2012) Morality-driven data forwarding with privacy preservation in mobile social network. IEEE Trans Veh Technol 61(7):3209–3222
11. Hao Y, Tang J, Cheng Y (2013) Secure cooperative data downloading in vehicular ad hoc networks. IEEE J Sel Areas Commun 31(9):523–537
12. Boneh D, Franklin M (2001) Identity-based encryption from the Weil pairing. In: Annual international cryptology conference, vol 2139, LNCS, Springer, pp 213–229
13. Soran H (2014) Lightweight security solutions for LTE/LTE-A networks. Networking and internet architecture [cs.NI]. Universit_e Paris Sud—Paris XI
14. Rukhin A, Soto J, Nechvatal J, Smid M, Barker E (2001) A statistical test suite for random and pseudorandom number generators for cryptographic applications

Efficient Data Collection Using Dynamic Mobile Sink in Wireless Sensor Network

Althiya Eby Irish, Sebastian Terence and Jude Immaculate

Abstract The elemental task of a wireless sensor network is to monitor a substantial area which requires gathering information by the sensor nodes from the sensor field and disseminate to the sink node. Data dissemination from cluster head to sink node in multiple hops leads to non-uniformity of energy consumption among sensors which may reduce network lifetime. Hence, the concept of sink mobility came into existence. For the complete utilization of sink mobility, a random move with efficient data collection is proposed which is named as Dynamic Sink Mobility for Data Collection (DSMDC). Sink migration is based on Detected Event Frequency (DEF). In DSMDC, the mobile sink plays the entire role for collecting data from cluster heads by first reaching the grid set with highest DEF and then by taking a clockwise move with SDMA technology to collect data from the remaining grid set in parallel while returning back to its original position. Existing data collection techniques that uses sink mobility are classified and analyzed. DSMDC is simulated using NS2 and proved to be better in terms of performance metrics when compared with two more dynamic moving strategies, namely DEF-A and DEF-D, in which the mobile sink moves based on the ascending and descending order of the DEF.

1 Introduction

A wireless sensor network (WSN) contains a wide range of sensor nodes that communicate wirelessly with each other [1]. Sensor network is used to monitor a larger area which requires gathering information by the sensor nodes from the sensor field and disseminate to the sink node. Data dissemination is the process of

A. E. Irish · S. Terence (✉)
Department of CST, Karunya Institute of Technology and Sciences, Coimbatore, India
e-mail: jsebinfo@gmail.com

J. Immaculate
Department of Mathematics, Karunya Institute of Technology and Sciences, Coimbatore, India

A. M. Zungeru et al. (eds.), *Wireless Communication Networks and Internet of Things*, Lecture Notes in Electrical Engineering 493,
https://doi.org/10.1007/978-981-10-8663-2_15

transmitting the gathered data to the sink from where the information can be obtained by the end users, and it is supposed to determine the optimal way from the source to the sink. Various algorithms and protocols are proposed for efficient data dissemination that can reduce or provide equalized consumption of energy on each sensor node in the network.

During data dissemination, the nodes which are closest to the static sink are expected to expend their battery supplies before other nodes due to heavy data traffic toward the sinks; such nodes are called as hotspots [2]. Either to eliminate or mitigate hotspots, the concept of mobile sinks was introduced. Multiple mobile sink technologies are also introduced recently to make the emergencies reported to the sink immediately. In this paper, Dynamic Sink Mobility for Data Collection (DSMDC) algorithm is proposed which effectively utilizes the sink mobility in a dynamic mode by moving based on the detection event frequency. Our objective is to make use of sink mobility to leverage traffic burden and thereby to diminish energy consumption and sustain network lifetime. Sink node will work like mobile base stations and collect data from associated cluster head sensors. Sink migration is based on the detection event frequency. The rest of the paper is structured as follows: Sect. 2 recollects the related work. Section 3 presents concept of the dynamic sink mobility algorithm, Sect. 4 evaluates algorithms, Sect. 5 concludes this paper.

2 Related Work

Data collection techniques are mainly classified into two categories based on the ability of sink movement—data collection using static sink and data collection using mobile sink. In stationary sink, sensor nodes transmit the collected data to the sink node. The data collected at each node can be transmitted in multiple hops through various nodes or through direct transmission to sink node. Various protocols that examine the merits and demerits of static sink are proposed [7, 8].

2.1 Data Collection Using Mobile Sink

We have classified the mobile sink category further into three divisions, namely mobility based on trajectory design, random sink mobility, and controlled sink mobility. In [3], the author contributed prediction-based data-aware clustering (PDC). It employs spatial and temporal correlations which help nodes in the cluster to sense similar values. By using local prediction models of sink nodes, cluster head forecasts readings in the network. In [5], the authors used Hilbert curve to re-dimension the mobile sink trajectory. Efficient network coverage and scalability are achieved, but the delivery delay is not minimized. In [9], the authors proposed optimal deadline-based trajectory (ODT) which helps to find out a trajectory for a mobile sink without the regard of any virtual structures in the network. As the mobile sink collects data from active sensor nodes in one-hop transmission, the network lifetime is sustained. In [4], the authors used multiple mobile sinks for data dissemination (MSD), which employs global agent to track sink locations. MSD constructs a two-tier grid structure and exploits hierarchical monitoring system in which sink mobility is random. In [6], the authors proposed biased random walk for the sink node of a wireless sensor network. The movement of the sink can be interfered by inherency of impediments. A random walk is proposed which uses probabilistic transitions between the cells.

From the survey, we found that sink mobility is necessary to improve network performance. The approaches that use sink mobility do not completely eliminate multi-hop data transmission in the network. Also, a predefined trajectory for the sink is possible only for theoretical analysis. Hence, a virtual grid network with random sink mobility that facilitates single-hop data transmission is proposed which reduces energy consumption and increases network lifetime. The performance of DSMDC is compared and proved to two more algorithms: DEF-A and DEF-D.

3 Dynamic Sink Mobility for Data Collection

3.1 System Model

A large-scale sensor network is considered where each node presumes either as source or router node. Source node is to sense data from the ambience. Router node is to transmit the observed data to the sink node. Each sensor node is assumed to be aware of its own geographical location using GPS. Sensor nodes communicate with cluster head in single-hop, and cluster head also communicates in single-hop with the sink. The proposed protocol is based on a virtual grid structure, where each cell represents a certain number of nodes. Cluster head selection is explained in Section B. Data collected by each sensor node is aggregated by its cluster head and transmitted to the sink using single-hop communication. Sink can move from one grid to another grid. Figure 1 shows the architecture of the proposed system.

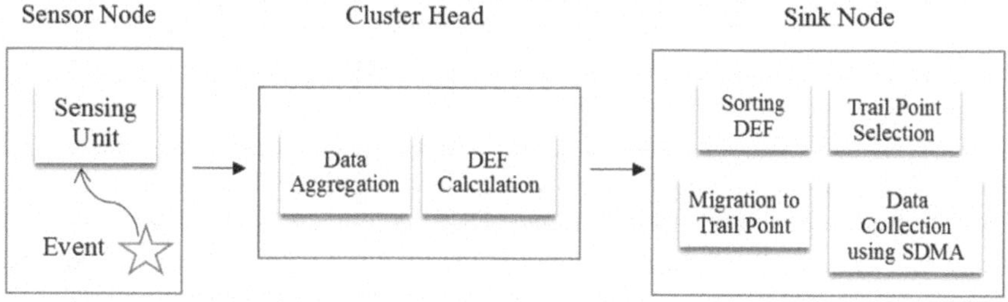

Fig. 1 System architecture

3.2 Cluster Head Selection

Once the network is partitioned into virtual grid, a cluster head for each grid is elected. Initially, each node of the network awaits receiving cluster head candidacy message. If that message is from the node of its own cluster, then it sets up the sender of the message as cluster head. The role of cluster head is twirled among sensors based on the residual energy of the currently elected cluster head. Power of sensor nodes are managed dynamically using two different threshold values, namely *Max_threshold* and *Min_threshold*. *Max_threshold* is the maximum energy available with a sensor node, whereas *Min_threshold* is the least residual energy. To select another node in the stage of energy exhaustion, *Min_threshold* is advertised. Whenever a node's energy reaches Min_threshold, cluster head selection is repeated.

3.3 Detection Event Frequency Calculation

Detection event frequency denotes the number of events detected per cluster during Mobility Time Period (MTP). Sink node maintains a table consisting of cluster ID and its appropriate detection event frequency. This information is sorted, and grid set which has the highest load of data is identified. Then the sink moves according to the proposed sink migration algorithm.

3.4 Sink Migration

Initially, sink node lies at the center of the complete gird structure. For each MTP, it gathers information from the clusters regarding the detected data. Sink node collects data in parallel from the cluster head using the standard SDMA technology. Three different sink migration algorithms, namely Dynamic Sink Mobility for Data Collection (DSMDC), Detected Event Frequency sorted in Ascending order (DEF-A), and Detected Event Frequency sorted in Descending (DEF-D), are proposed and analyzed. The result shows that DSMDC consumed less energy with improved network lifetime, and throughput of DSMDC is higher than other proposed technique.

DSMDC takes the input of the DEF from various grid sets and returns trail points (TPs), which should be followed by the mobile sink in data collection. DSMDC finds out the grid set with highest DEF and that grid set will be visited first by the mobile sink. Visited grid set will be the first trail point. The rest of the trail points are also found in the steps 8–12 in Algorithm 1, where the mobile sink is subjected to proceed in a clockwise route among the unvisited grid sets. Trail pointer algorithm takes in the cluster coordinates of a grid which has highest DEF and finds out the position where the sink node is supposed to move next. DEF-A and DEF-D sort the detection events of each grid in ascending and descending order, respectively, and allows the sink to move in the order of the detected event frequency.

Algorithm 1: DSMDC

```
Input: Max_DEF
Output: Sink Migration Path
n ← Maximum number of grid set
D = { D1, D2, ... Dg }      /* DEF of each grid set g */
Trail Points = { TP1, TP2 ... TPg }
for each g in grid set
if Dg == Max_DEF
        TP1 ← g
        i←g
j←2
while j <= n
            TPj ← (i+1) mod n
            j ←j+1
            i ←i+1
return Trail Points;
```

Algorithm 2 Trail Point Finder

```
Input: DEF_grid (Cluster co-ordinates of a grid)
Output: Current_TrailPoint
IF x.DEF_grid <= 1
IF y.DEF_grid <=1
        Current_TrailPoint = TP_1
ELSE
        Current_TrailPoint = TP_2
ELSE
IF y.DEF_grid <=1
        Current_TrailPoint = TP_3
ELSE
            Current_TrailPoint = TP_4
END IF
return Current_TrailPoint
```

The grid structure is divided into four trail points. As mentioned earlier, initially sink node lies at center of the grid structure. The trail point finder algorithm is used to find out the position where the sink node is supposed to move next by using received DEF. DEF-A sorts the detection events of each grid in ascending order and DEF-D sorts the detection events of each grid in descending order. It allows the sink to move in the order of the detected event frequency.

Algorithm 3 DEF-A

```
Input: D = { D1, D2,...Dn}       /* DEF of each grid*/
Output: Trail Points
n = number of grids
A = sorted array of D in Ascending order
i = 0
while i < n
TP = Trail Point Finder (DEF_grid)
i←i+1
```

4 Result and Discussion

The experiment is conducted by simulation of Network Simulator-2 with sample of 100 nodes which are divided into four trail points and further trail point is split into virtual grid structure. The performance evaluation is calculated by metrics such as energy consumption, detection event frequency, throughput, response time, and packet delivery ratio. The simulation results of DSMDC are compared with two more dynamic mobility schemes, namely DEF-A and DEF-D, in the same simulator. The result depicts that DSMDC consumed less energy and increased network lifetime than other two schemes.

The energy E is calculated as the sum of the energy used by each sensor node to transmit, receive, and also the sum of energy expelled while carrier sensing and in sleep mode. Figure 2 shows that DSMDC consumes very less energy when compared to DEF-A and DEF-D. Density is the number of sensor nodes in a particular grid set. When the density increases, DEF-D consumes greater amount of energy, whereas DEF-A and DSMDC decrease in energy consumption.

Throughput in sensor networks is the average of successful delivery of data. Figure 3 shows that the throughput of DSMDC is higher than that of DEF-A and DEF-D. Throughput is the size of the data divided by time required to transmit that data. Packet delivery ratio is the ratio of data packets successfully received to that of sent.

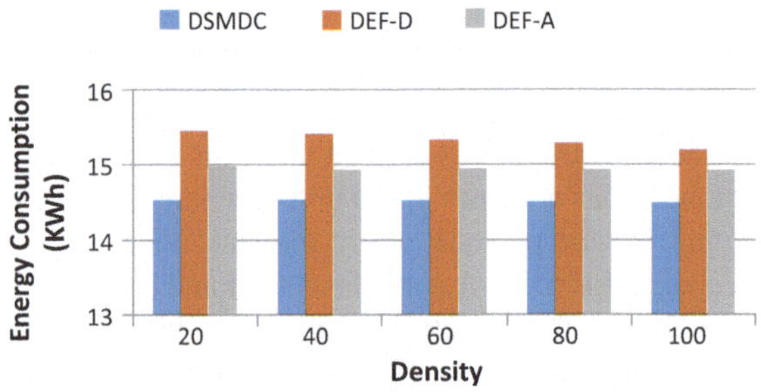

Fig. 2 Density versus energy consumption

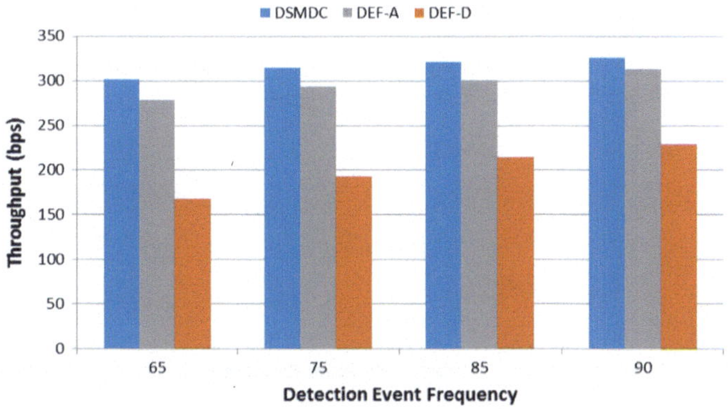

Fig. 3 Detection event frequency versus throughput

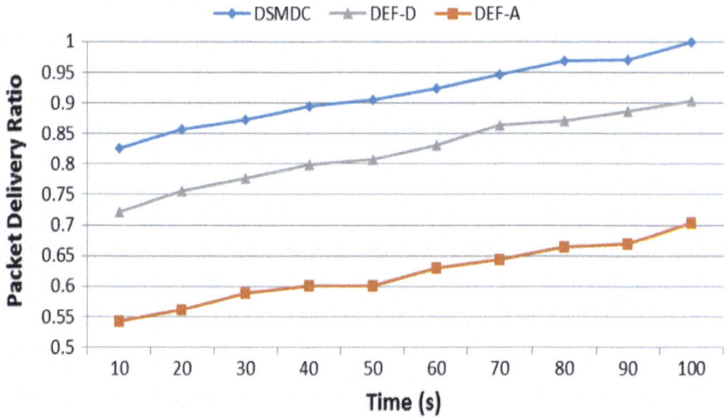

Fig. 4 Time versus packet delivery ratio

Response time is the time taken between the sending and receiving of packets. Response time and packet delivery ratio are compared with DEF which in Fig. 4 proves that regardless of DEF, DSMDC performs well. Hence, the best response time is given by DSMDC. Figure 5 depicts the packet delivery ratio of the DSMDC, which again outperforms the other two algorithms.

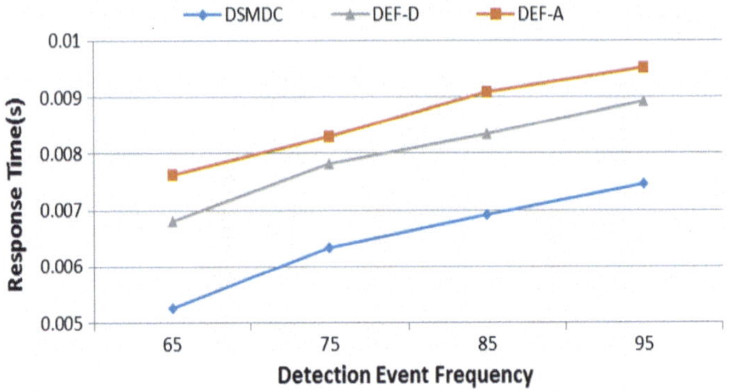

Fig. 5 Detection frequency versus response time

5 Conclusion

Sensor networks are known for their nature of having the limited lifetime of the sensors in phenomenon with their limited size, and their performances are based on the mode of communication. The proposed protocol is based on a structure of a virtual grid. Each cell in the grid contains a cluster head obligated on the collection, aggregation, and dissemination of the sensed data to the sink. This head is elected recurrently at regular intervals according to the residual energy of the sensors using dynamic power management. Three different sink migration strategies are proposed which avoid multiple-hop communication between cluster heads of the network. Sink node uses SDMA technology for parallel data collection. On evaluating performance of DSMDC comparing with other strategies, namely DEF-A and DEF-D, DSMDC is proved to consume less energy with improved network lifetime.

References

1. Akyildiz I, Su W, Sankarasubramanian Y, Cayirci E (2002) A survey on sensor networks. IEEE Commun Mag 40:102–114
2. Tunca C, Isik S, Donmez MY, Ersoy C (2014) Distributed mobile sink routing for wireless sensor networks: a survey. IEEE Commun Surv Tutorials 16:877–897
3. Ashouri M, Yousefi H, Basiri J, Hemmatyar AMA, Movaghar A (2015) PDC: prediction-based data-aware clustering in wireless sensor networks. J Parallel Distrib Comput 81–82:24–35
4. Dongliang XIE, Xiaojie WU, Dan LI, Jia S (2014) Multiple mobile sinks data dissemination mechanism for large scale wireless sensor network. China Commun 13:1–8
5. Ghafoor S, Rehmani MH, Cho S, Park SH (2014) An efficient trajectory design for mobile sink in a wireless sensor network. Comput Electr Eng 40:2089–2100
6. Kinalis A, Nikoletseas S, Patroumpa D, Rolim J (2014) Biased sink mobility with adaptive stop times for low latency data collection in sensor networks. Inf Fusion 15:56–63

7. Zhao Z, Dong W, Jiajun B, Yu G, Chen C (2015) Link correlation aware data dissemination in wireless sensor networks. IEEE Trans Industr Electron 62:5747–5757
8. Zheng X, Wang J, Dong W, He Y, Liu Y (2015) Bulk data dissemination in wireless sensor networks: analysis, implications and improvement. IEEE Trans Comput 65:1428–1439
9. Tashtarian F, Moghaddam MHY, Sohraby K, Effati S (2014) ODT: optimal deadline-based trajectory for mobile sinks in WSN: a decision tree and dynamic programming approach. Comput Netw 77:128–143

Dependency Analysis of Control Parameter Configuration on ISD and Random Mobility of UE in LTE-A Network

A. Saraswathi Priyadharshini and P. T. V. Bhuvaneswari

Abstract The adoption of UE-assisted, network-controlled, and hard handover (HO) procedure in LTE-A HetNet imposes several challenges such as performance degradation, complexity in network planning and maintenance. To alleviate these challenges, proper configuration of HO control parameters is needed to trigger the HO procedure in time. The control parameters need to be configured in consideration with various characteristics such as environmental, network, and user equipment (UE). Further, in the literature, analyzing the impact of configuring all the control parameters remains an open issue. Hence, in this work, the impact of control parameters involved in the Macro-Macro environment such as Hysteresis Margin, A3Offset, and Time-To-Trigger have been analyzed. The environmental impact has been analyzed both in urban and in rural scenarios. Also, the dependency of these control parameters with respect to the inter-site distance (ISD) between the base stations and UE mobility pattern is investigated.

Keywords LTE-A · HetNet · Handover · Control parameters · Inter-site distance

1 Introduction

In Long-Term Evolution-Advanced (LTE-A), the handover (HO) procedure suggested by 3GPP is UE-assisted, network-controlled, and Hard HO [1]. However, the HO mechanism needs to be initiated on time to maintain the ongoing session and also to provide user the better Quality of Service (QoS). The transmission time of measurement report (MR) in the HO procedure plays a vital role in making the triggered HO successful. Periodical or predefined event-based transmission of MR can be made [2].

A. Saraswathi Priyadharshini (✉) · P. T. V. Bhuvaneswari
Department of Electronics Engineering, MIT Campus, Anna University, Chennai, India
e-mail: ajssara@gmail.com

P. T. V. Bhuvaneswari
e-mail: ptvbmit@annauniv.edu

A. M. Zungeru et al. (eds.), *Wireless Communication Networks and Internet of Things*, Lecture Notes in Electrical Engineering 493,
https://doi.org/10.1007/978-981-10-8663-2_16

The network operators prefer mostly the event-based transmission of MR in order to reduce the consumption of resources. There are many events defined by 3GPP depending on the type of HO. Among those, A3 event is most predominantly utilized in the literature as it compares the signal strength of serving eNodeB against the target eNodeB. The transmission of MR is made whenever the UE experiences the A3 event condition for Time-To-Trigger (TTT) duration; otherwise, it is not transmitted. Hence, the control parameters of A3 event such as *Hysteresis Margin (HM), A3Offset (A3Off)* as well as TTT decide the triggering time of HO.

However, it is observed from the literature survey that the impact of all the control parameters in varied network and environmental and user characteristics has not yet been analyzed. Hence, in this work, analysis has been made to study the impact of network characteristic called inter-site distance (ISD) as well as the impact in rural and urban environment with random mobility of UE. The combined impact of control parameters with respect to ISD, UE movement, and TTT on HO performance has been investigated.

The organization of the paper is as follows: Sect. 2 discusses the literature related to HO control parameter configuration. Section 3 discuss the performance analysis. Results and inference of the proposed model are elaborated in Sect. 4. Finally, the paper is concluded with future directions in Sect. 5.

2 Literature Survey

This section details the state of the art in the configuration of HO control parameters. Authors in [3] have developed adaptive and grouping techniques by analyzing the impact of HM and TTT control parameter on HO performance in terms of radio link failure (RLF) and ping-pong (PP) rate. From the simulation results obtained, it is inferred that adaptively selected TTT value results in better HO performance and also, the inclusion of HM mitigates the effect of ping-pong rate to a maximum extent.

The authors in [4] have analyzed the impact of control parameters such as TTT, A3Off, and cell-specific offset (CSO) in Macro-Pico HO scenario. The dependency of these parameters on HO performance has been investigated in detail. It has been concluded that the HO performance can be improved when these parameters are configured considering the mobility pattern of UE. The authors in [5] have studied the reliance of the metric ISD between Macro- and Femto-cell on HO performance. Results show that RLF and PP probability increases with increase in ISD and vice versa. It is concluded that the HO performance can be improved only when the value of TTT is selected based on ISD, UE profile, and environmental characteristics.

In [6], authors have analyzed the impact of control parameters such as HM, A3Off, and TTT in Macro-only scenario. The analysis has been made in environment of different ISDs. From the simulation results, it is inferred that the same combination of control parameter which triggers the A3 event too early for smaller ISD triggers the A3 event too late for larger ISD. However, this work analyzes the straightline movement of UE from serving to target base station.

From the above literature survey, it has been observed that the configuration of control parameters plays a significant role in improvising the HO performance. However, most of the literature lacks in analyzing the combined impact of all the parameters and their influence on deployment characteristics. Hence, an attempt has been made in this research to study the interdependency of the control parameters with respect to ISD and their combined impact on HO performance. Further, the impact of straightline and random mobility of UE on HO performance has been analyzed.

3 Proposed Handover Performance Analysis

This section details the steps involved in the proposed HO performance analysis.

3.1 System Model

Let SMeNB and TMeNB be the Serving and Target Macro eNodeB placed with the inter-site distance of ISD_{MM} (Macro-Macro), and the UEs move with a velocity 'V_i' making straightline and random movements.

3.2 Proposed Analysis

Figure 1 represents the steps involved in the proposed HO performance analysis. Initially, the control parameters such as *HM*, *A3Off*, and *TTT* involved in Macro-Macro (MM) HO scenario analysis are configured. Then, the occurrence of A3 event and persistence of it for the duration of TTT are verified for initiating the transmission of MR.

The A3 event trigger distance (D_{A3T}) is the distance from SMeNB at which MR is transmitted during UE's mobility and is defined in Eq. (1).

$$D_{A3T} = M_T - HM > M_S + A3Off \tag{1}$$

where M_S and M_T are the reference signal received power (RSRP) of serving and target base station measured in dBm. Upon triggering the MR, it is verified whether the initiated HO results in success or failure. It is done by verifying three conditions such as (i) successful reception of MR by SMeNB, (ii) successful reception of HO command by UE after HO preparation delay, and (iii) persistence of radio link with

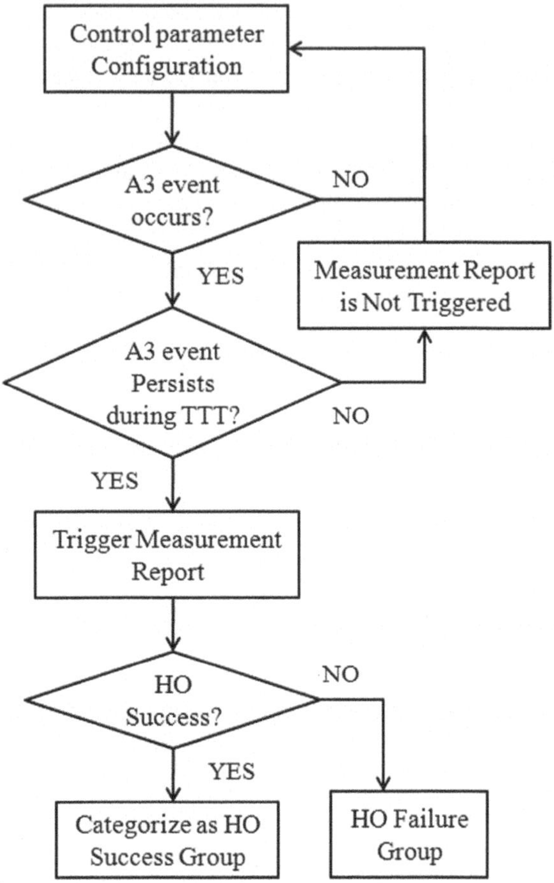

Fig. 1 HO performance analysis

TMeNB after HO execution delay [3]. The control parameters are reconfigured if A3 event does not occur and also when it fails during TTT. This procedure is continued for all the combinations of HO control parameters.

3.3 Impact Analysis

The impact of HO control parameter on HO performance is analyzed in an urban environment with UEs making a straightline as well as random movement. In this work, the angle formed by the UEs (α_i) moving in straight line is considered to be of 10°, 30°, 50°, and 70°. Initially, the number of groups which initiates HO success for different angles and ISD_{MM} varying from 500 to 1100 m has been analyzed. And the impact of TTT and ISD_{MM} in rural and urban environment on HO performance has been investigated. The ISD_{MM} in urban environment is considered to be 500, 700, 900, and 1100 m, while in rural environment the values of ISD_{MM} considered are 1500, 1700, 1900, and 2100 m.

4 Results and Discussion

The presented work is simulated in MATLAB simulator version R2014a. The results of the analysis made in MM HO scenario are discussed in this section. The simulation parameters and their associated values are mentioned in Table 1.

4.1 Group Formulation

The range of HM and A3Off values considered in this research is as per 3GPP specification [2], and 28 combinations of control parameters involving HM and A3Off are formulated. From the analysis, the combinations of control parameters which result in the same value of D_{A3T} are grouped. The resulting 10 groups are from G1 to G10 as mentioned in Table 2. The individual groups along with respective TTT are then configured to investigate the impact on HO performance. The groups from G1 to G10 which satisfy the three conditions mentioned in previous section are referred to as HO success groups.

Table 1 Simulation parameters

Parameter	Configuration
Transmitter power	Macro eNodeB: 46 dBm
Propagation model—urban	128.31 + 37.06 (log(*R*)), '*R*' in Km
Propagation model—rural	96.31 + 34.07 (log(*R*)), '*R*' in Km
UE transmit power	30 dBm
Time-to-trigger	{40, 1280, 5120 ms}
Hysteresis margin	{0, 5, 10, 15 dB}
A3Offset	{−15, −10, −5, 0, 5, 10, 15 dB}
HO delay	Preparation time: 50 ms, execution time: 40 ms
Velocity of UE (V)	20, 50 and 110 km/h
Radio link failure	RSRP should be less than −130 dBm
Minimum required RSRP ($RSRP_{min}$)	−110 dBm (3GPP TS 36.301)
Antenna type	Omnidirectional antenna
Antenna gain	eNodeB: 15 dBi, UE: 0 dBi
Mobility model	Random waypoint mobility

Table 2 Groups formulated with same $A3_{TD}$

Groups	(HM, A3Off) in dB
G1	(0, −15)
G2	(5, −15) (0, −10)
G3	(10, −15) (5,−10) (0, −5)
G4	(15, −15) (10, −10) (5, −5) (0, 0)
G5	(15, −10) (10, −5) (5, 0) (0, 5)
G6	(15, −5) (10, 0) (5, 5) (0, 10)
G7	(15, 0) (0, 15) (10, 5) (5, 10)
G8	(5, 15) (10, 10) (15, 5)
G9	(15, 10) (10, 15)
G10	(15, 15)

4.2 HO Simulation Scenario

Figure 2 shows the MATLAB simulation scenario, where the triangle symbol (Δ) represents the base stations distributed with ISD_{MM} and UEs moving in random nature.

4.3 Impact of UE Mobility Pattern and ISD_{MM} on the Performance of HO

The impact of UE mobility pattern of straightline and random movements on HO performance when configured with G1 to G10 groups is analyzed under the fixed condition of $ISD_{MM} = 500, 700, 900$, and 1100 m; $V_{max} = 110$ km/h; and $TTT_{max} = 5120$ ms. Figure 3 represents the number of HO success groups for two

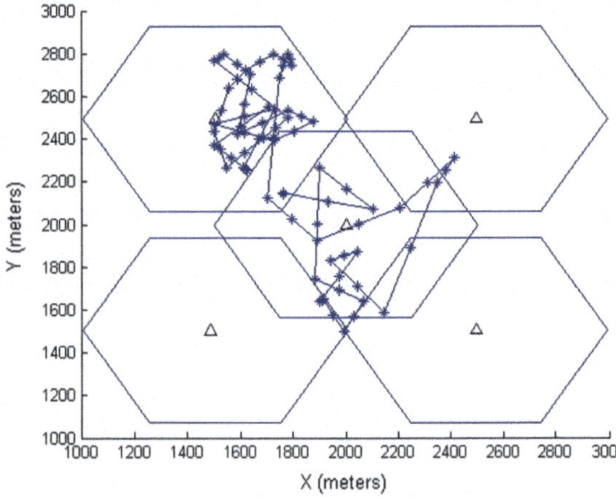

Fig. 2 Simulation scenario of UE's random mobility

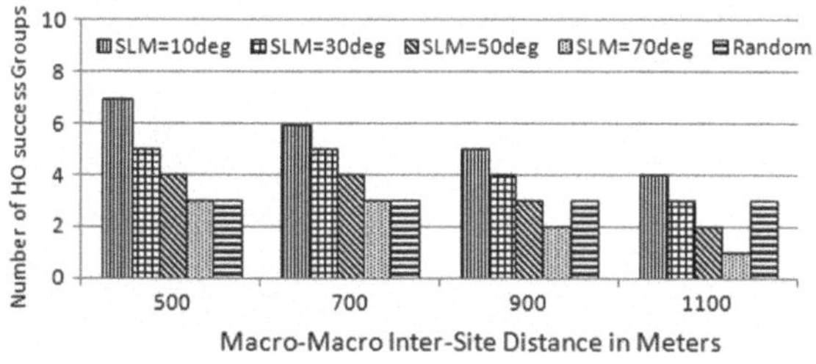

Fig. 3 HO triggering distance for different ISD_{MM}

scenarios, namely (i) straightline movement (SLM) of UE making angle α_i = 10°, 30°, 50°, and 70° and (ii) random movement.

From the results, it is inferred that as the value of 'α_i' and ISD_{MM} increases, there is decrease in the number of HO success groups. This is because late triggering of HO when the UE is moving farther away from the target eNodeB results in HO failure. Also, early triggered HO results in success for random movement of UE. The results ensure that success of HO can be attained only when it is triggered early irrespective of UE movement and ISD_{MM}.

4.4 Impact of ISD_{MM} and TTT on HO Success in Urban Environment

The impact of ISD_{MM} and TTT on HO performance when configured with groups from G1 to G10 is analyzed in urban environment. The values of ISD_{MM} considered are 500, 700, 900, and 1100 m and random movement with velocity of 110 km/h. Figure 4 represents the number of HO success groups when configured with TTT values of 40, 1280, and 5120 ms.

From the results, it is observed that for ISD_{MM} of 500 m the group G4 results in HO success. Further, for different ISD_{MM} configuring TTT has no impact on HO success groups. The early triggering groups such as G1, G2, and G3 result in HO success irrespective of TTT and ISD_{MM}. This is because the completion of HO procedure occurs before the radio link failure as it is triggered early.

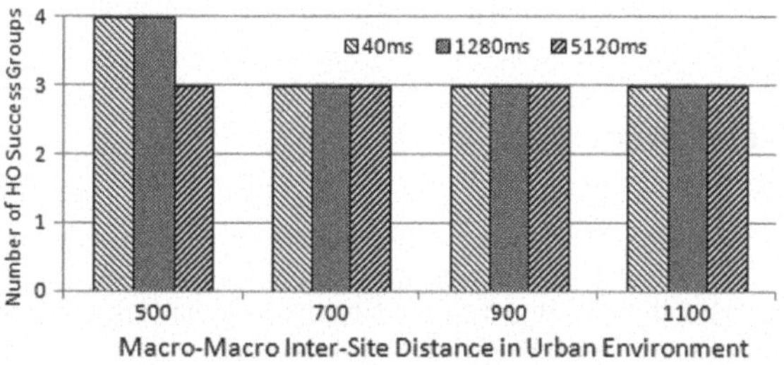

Fig. 4 HO performance in urban environment

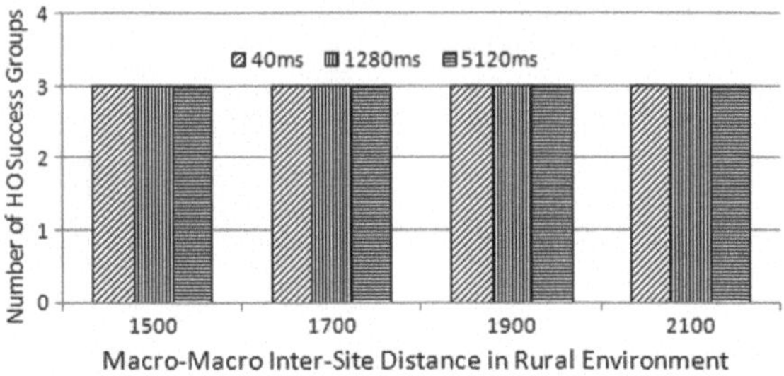

Fig. 5 HO performance in rural environment

4.5 *Impact of ISD_{MM} and TTT on HO Success in Rural Environment*

Similarly, the above analysis is made in rural environment and represented the outcome in Fig. 5. The values of ISD_{MM} considered are 1500, 1700, 1900, and 2100 m, and UEs move randomly with velocity of 210 km/h. From the results, it is observed that for all the considered ISD_{MM} and TTT of 40, 1280, and 5120 ms, the groups from G1 to G3 result in HO success. This is because these groups trigger HO early and get completed before the radio link failure. Hence, the groups which trigger HO early result in success irrespective of TTT and ISD_{MM}.

5 Conclusion and Future Work

In this work, the combined impact of control parameters such as HM, A3Off, and TTT on HO performance in Macro-Macro environment has been investigated with respect to ISD_{MM} and UE mobility pattern. The analysis of random moving UE in urban and rural environment has been made to study the impact of TTT and ISD_{MM}. The results strongly emphasize that configuration of above-mentioned control parameters completely depend on these factors with straightline movement of UE. However, the control parameters have a negligible dependency on ISD_{MM} and TTT with random movement of UE. In future, the analysis has to be extended to HetNet scenario of Macro-Pico deployment by taking into account the QoS characteristics of UEs also. Further, from the conclusions of this analysis, reinforcement learning algorithm has to be formulated.

References

1. Kottkamp M, Roessler A, Schlienz J (2012) LTE-advanced technology introduction. White paper. Rohde & Schwarz, pp 1–41
2. 3GPP TS 36.331, V13.1.0 (2016) Evolved universal terrestrial radio access (E-UTRA); radio resource control (RRC); protocol specification

3. Lim J, Hong D (2013) Mobility and handover management for heterogeneous networks in LTE-advanced. J Wirel Pers Commun Springer 72:2901–2912
4. Mehta M, Akhtar N, Karandikar A (2015) Impact of handover parameters on mobility performance in LTE HetNets. In: Twenty-first national conference on communications, pp 1–6
5. Kollias G, Adelantado F, Verikoukis C (2015) The impact of inter-site distance and time-to-trigger on handover performance in LTE-A HetNets. In: International conference on communications, pp 3969–3974
6. Saraswathi Priyadharshini A, Bhuvaneswari PTV (2016) A study on handover parameter optimization in LTE-a networks. In: International conference on microelectronics, computing and communication pp 1–5

Throughput Analysis of MacroUE for Varied Transmit Power of Small Cell in Heterogeneous Network

S. Ezhilarasi and P. T. V. Bhuvaneswari

Abstract Femtocells are the solution introduced in long-term evolution-advanced (LTE-A) network to enhance the data rate and coverage of indoor users. The deployment of femtocell over macrocell is known as heterogeneous networks (HetNets). The available spectrum is shared between macrocell and femtocell in order to improve the system performance. However, the unplanned deployment of femtocell causes severe interference to macrocell and vice versa. The interference can be either co-tier or cross-tier. The objective of this research is to mitigate cross-tier interference experienced by Macro User Equipment (MUE) when placed within the vicinity of femtocell. The performance of the proposed analysis is examined in terms of signal-to-interference-plus-noise ratio (SINR) and throughput. From the simulation results, it is found that as the distance between MUE and macrocell decreases, throughput and SINR experienced by the MUE increase and vice versa.

Keywords Femtocell · HetNet · MUE · SINR and throughput

1 Introduction

HetNet feature of LTE-A improves the coverage and capacity of indoor users through deployment of small cells over macrocell [1]. Out of the small cells, picocells are specified by network operator, while femtocells are planned by users. Hence, challenges imposed by femtocells need to be given more attention when compared to picocell. Deployment of these small cells results in network performance degradation which is caused due to cross-tier and co-tier interference.

S. Ezhilarasi (✉) · P. T. V. Bhuvaneswari
Department of Electronics Engineering, MIT Campus, Anna University, Chennai, India
e-mail: ezhilvish@yahoo.co.in

P. T. V. Bhuvaneswari
e-mail: ptvbmit@annauniv.edu

A. M. Zungeru et al. (eds.), *Wireless Communication Networks and Internet of Things*, Lecture Notes in Electrical Engineering 493,
https://doi.org/10.1007/978-981-10-8663-2_17

The co-tier interference occurs between neighboring femtocells, while cross-tier interference occurs between macro- and femtocells in the network [2].

The cross-tier interference experienced by UEs of both macrobase station (MBS) and femtobase station (FBS) depends upon two factors, namely distance between MBS-FBS, MUE-MBS, and MUE-FBS. Further, the transmission power of FBS also influences the level of interference. Hence, the main objective of this research is to analyze the impact of transmission power of FBS and their placement on throughput of MUE served by MBS.

The remaining of the paper is organized as follows: Sect. 2 details the state of art related to the considered research. Section 3 presents the system model considered for the investigation. Section 4 describes the performance of the carried out research along with results. Finally, Sect. 5 concludes the paper with future work.

2 Literature Survey

Various combinations of HetNet prescribed by 3GPP in LTE-A are macro-pico, macro-femto, and macro-micro. Macro-femto has more implementation challenges because their deployment is planned by users. The unplanned deployment of FBS causes severe cross-tier interference when compared to co-tier interference. In this research, the analysis is confined to cross-tier interference in downlink macro-femto scenario. Several solutions exist in the literature that addresses the downlink cross-tier interference [3–7].

In [3], the authors have analyzed adaptive coverage of FBS to mitigate the downlink cross-tier interference. It is inferred that enhancement in spectral efficiency and throughput is achieved by COST-Hata and ITU indoor propagation model.

In [4], authors have analyzed the impact of cross-tier interference experienced by MUE in both indoor and outdoor environment. However, the results of outdoor environment are alone discussed.

In [5], the authors have analyzed the dependency of load and transmit power of FBS on SINR of MUE in a macrourban microscenario. It is found that as level of interference experienced by MUE due to FBS increases, degradation can be witnessed.

In [6], the authors have derived closed expression for SINR using analytical fluid model. The conclusion drawn is QoS of MUE in a HetNet scenario which depends on the impact of deployment of FBS, their configuration, and distance with respect to UE in addition to their transmitting power.

In [7], the authors have addressed the downlink interference problem by priced water filling and adaptive spectral mask algorithms. The performances of both the algorithms are analyzed in terms of interference experienced by MUE.

From the above literature, it is found that adaptive power control scheme is one of the solutions to mitigate downlink cross-tier interference in macro-femto scenario. In this paper, an attempt is made to analyze the throughput of MUE in both indoor and outdoor environment in consideration with path loss (PL) models. Further, the impact of transmission power of FBS and their placement along with modulation schemes are investigated.

3 System Model

In this section, the system model executed in the proposed research is presented.

3.1 Downlink Cross-Tier Interference Model

The FBS is assumed to be operated in closed access mode which is modeled as two-dimensional house with single floor. Let S be the spectrum shared between MBS and FBS, Δf denotes the subcarrier spacing, $M = \{1,\ldots,m\}$, $F = \{1,\ldots f\}$, $U = \{1,\ldots u\}$ be the numbers of MBS, FBS, and UEs are served by MBS, respectively. Then, $d_{i,j}$, $d_{j,k}$, and $d_{i,k}$ are the distance between MBS-MUE, MUE-FBS, and FBS-MBS, respectively, as shown in Fig. 1. Let P_M, $P_F = \{1,\ldots p\}$, P_U be the transmit power of MBS, FBS, and MUE, respectively, such that $\{P_M > P_U > P_F\}$. Let P_r be the reference signal received power (RSRP) of MUE with respect to MBS, P_i be the interference power due to FBS, and $\mathrm{PL_{UM}}$ and $\mathrm{PL_{UF}}$ be the PL between MUE and MBS, MUE and its interfering FBS, respectively.

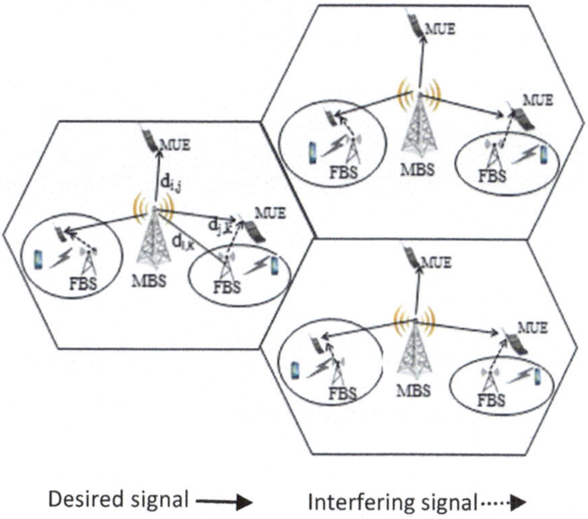

Fig. 1 System model of the proposed analysis

3.2 Femto-Macro Interference

The proposed research analyzes the interference experienced by MUE served by MBS due to the presence of FBS. It consists of various stages such as

(i) Deployment of base station and user equipment
(ii) Computation of distance
(iii) Computation of path loss
(iv) Computation of throughput of MUE

3.2.1 Deployment of Base Station and User Equipment

The deployment of base station formulates a two tier network in which FBS is overlaid within the MBS. In this research, the MUE served by MBS and FBS deployed in indoor deployment are considered. Further, the impacts of wall penetration are alone considered in the analysis.

3.2.2 Computation of Distance

Let $(X_i\ Y_i)$, (X_j, Y_j) be the coordinates of MBS and MUE. Then, $d_{i,j}$ represents the Euclidian distance between MUE and MBS. It is computed using expression (1)

$$d_{i,j} = \sqrt{(Y_j - Y_i)^2 + (X_j - X_i)^2} \tag{1}$$

Let (X_k, Y_k) be the coordinates of FBS, then Euclidian distance $d_{j,k}$ is computed between FBS and MUE using Eq. (2)

$$d_{j,k} = \sqrt{(Y_k - Y_j)^2 + (X_k - X_j)^2} \tag{2}$$

Similarly, the Euclidian distance $d_{i,k}$ between MBS and FBS is found using Eq. (3)

$$d_{i,k} = \sqrt{(Y_k - Y_i)^2 + (X_k - X_i)^2} \tag{3}$$

3.2.3 Calculation of Path Loss

In a wireless channel, the medium between transmitter and receiver is highly influenced by multipath components such as PL and shadowing which causes degradation to the transmitted signal. Several PL models are specified in 3GPP [8].

In this work, the influence of PL is alone considered of Macro-femto scenario in suburban environment [9]. The PL between MUE (indoor and outdoor) and its serving MBS is computed using the model in (4)

$$\mathrm{PL}_{\mathrm{UM}}(\mathrm{db}) = 15.3 + 37.6\log_{10} d_{i,j} + L_{\mathrm{ow}} \tag{4}$$

where '$d_{i,j}$' represents the distance between MUE and MBS and it is in 'm,' and L_{ow} is the wall penetration loss that takes into account in indoor environment. Its value can be either 10 dB or 20 dB. Similarly, PL is between indoor MUE and FBS and outdoor MUE and FBS are computed using model (5) and (6), respectively.

$$\mathrm{PL}_{\mathrm{UF(indoor)}} = 38.46 + 20\log_{10} d_{j,k} + 0.7_{d_{2d,\mathrm{indoor}}} + 18.3._{n}^{(n+2/n-1)-0.46} \tag{5}$$

$$\begin{aligned} PL_{\mathrm{UF(outdoor)}} = \max(&15.3 + 37.6\log_{10} d_{i,j}, 38.46 + 20\log_{10} d_{j,k} \\ &+ 0.7_{d_{2d,\mathrm{indoor}}} + 18.3._{n}^{(n+2/n-1)-0.46} + L_{\mathrm{OW}} \end{aligned} \tag{6}$$

where 'n' is the number of penetrating floors and d_{2d} indoor represents MUE distance inside the house in 'm.'

3.2.4 Computation of Throughput

The performance of MUE is determined based on the throughput. This in turn is calculated from SINR. Let 'k' be the subcarrier allocated for both MBS and FBS, then the SINR of MUE is computed using expression (7).

$$\mathrm{SINR}_{U,K} = \frac{P_{M,K} G_{U,M,K}}{N_o \Delta_f + \sum_M P'_{M,K} G_{U,M',K} + \sum_F P'_{F,K} G_{U,P,K}} \tag{7}$$

where $P_{M,k}$ and $P'_{M',k}$ are the transmitting power of serving and neighboring MBS on subcarrier k and $G_{U,M,k}$ and $G_{U,M',k}$ are the channel gain between MUE with serving and neighboring MBS on subcarrier 'k.' $P'_{F,k}$ is the transmit power of neighboring FBS on subcarrier k, $G_{U,p,k}$ is the channel gain between UE and its serving or interfering BS on subcarrier k, N_o denotes the noise power spectral density, and Δ_f denotes the subcarrier spacing.

The channel gain (G) is computed from PL using the following expression.

$$G = 10^{-\mathrm{PL}/10} \tag{8}$$

where $G = G_{U,M,k}, G_{U,M',k}, G_{U,p,k}$.

Then, the throughput of MUE served by MBS on subcarrier 'k' is computed using Eq. (9).

$$C_{u,k} = \Delta f \,.\, \log_2^{(1 + a\mathrm{SINR}_{U,K})} \tag{9}$$

where $a = -1.5/\ln(5\mathrm{BER})$ and BER represents bit error rate.

4 Results and Discussion

The simulation parameters considered are as per 3GPP. It is presented in Table 1.

The HetNet consist of one center excited MBS of radius 250 m. FBS with radius of 10 m is deployed at three different positions, namely near, middle, and edge from the MBS.

4.1 SINR Analysis

The interference experienced by MUE is analyzed in terms of SINR and throughput when MUE is located at 2 m distance from FBS.

Figure 2 shows the SINR of indoor MUE located at three different positions within the vicinity of FBS. This implies that the received power is analyzed with respect to MUE path loss between serving MBS and its interfering FBS by their deployment. Further, the impact of SINR is examined based on FBS transmit power. It is inferred that SINR is increased by 4% for every 5 dbm transmitting power of FBS with respect to its maximum power of 20 dbm when MUE is located near to its serving MBS.

Table 1 Simulation parameters

Parameter	Value
Macrocell radius (R_m)	250 m
Femtocell radius (R_f)	10 m
Carrier frequency (f_c)	2.5 GHz
Macrocell transmit power (P_M)	43 dBm
Femtocell transmit power (P_F)	10 dBm, 15 dBm, 20 dBm
MUE transmit power (P_U)	23 dBm
Bandwidth (BW)	5 MHz
Minimum distance between MBS and FBS ($d_{i,k}$)	35 m
Minimum distance between MUE and FBS ($d_{j,k}$)	<0.2 m
Number of active MUEs	10
Modulation scheme	16QAM, 64QAM
Wall penetration loss (L_{ow})	10 dB
Subcarrier spacing (Δ_f)	15 kHz
White noise power density (N_o)	−174 dBm/Hz

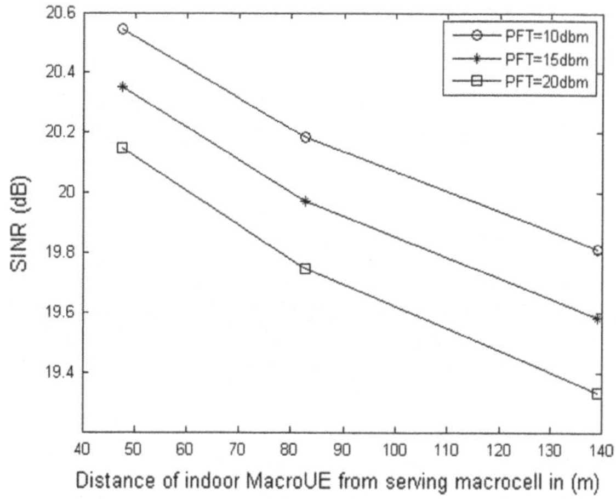

Fig. 2 SINR of indoor MUE accounting interference from FBS

Next, the same approach is followed for outdoor MUE, since the wall penetration loss is excluded between outdoor MUE and MBS. From Fig. 3, it is found that SINR is increased by 26% for 10 dbm transmitting power of FBS compared to that of indoor environment. It is inferred that SINR is improved due to propagation effects in the channel with respect to MUE distance from the serving MBS. Similarly, the MUE located at distance 5 and 10 m from FBS is also examined which reveals that received power varied due to MUE-FBS distance.

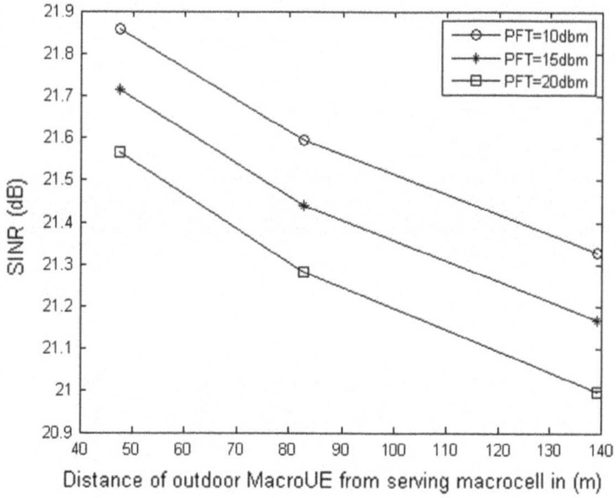

Fig. 3 SINR of outdoor MUE accounting interference from FBS

4.2 Throughput Analysis

From Fig. 4, it is noted that MUE obtained 6.4% improvement in throughput for every 5 dbm transmitting power of FBS using 16-quadrature amplitude modulation (QAM).

From Fig. 5, throughput is increased by 27% using 64-QAM compared to that of 16-QAM when MUE is located near to FBS with the transmitting power of 10 dbm. The same observations are carried out in the case of outdoor environment without considering wall penetration loss as shown in Figs. 6 and 7.

Fig. 4 Throughput of indoor MUE using 16-QAM

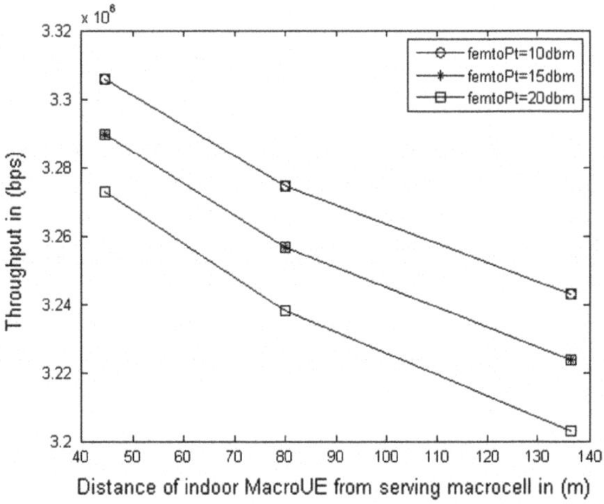

Fig. 5 Throughput of indoor MUE using 64-QAM

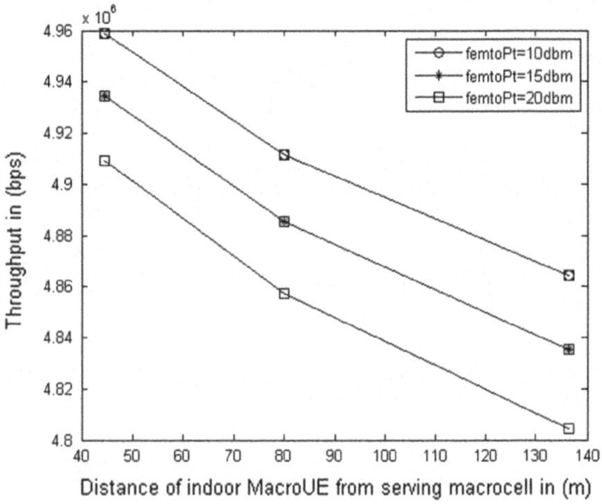

Fig. 6 Throughput of outdoor MUE using 16-QAM

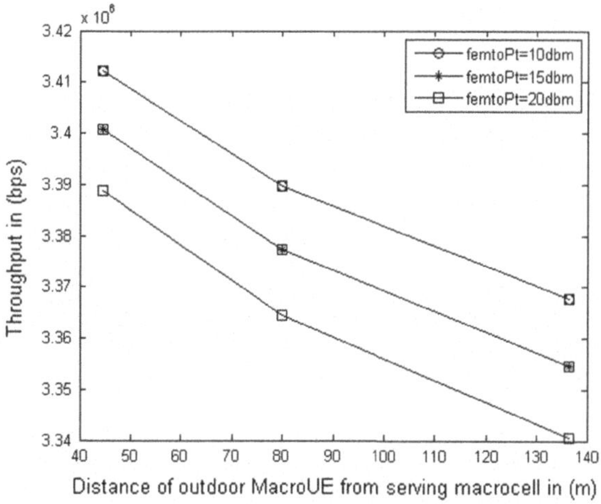

Fig. 7 Throughput of outdoor MUE using 64-QAM

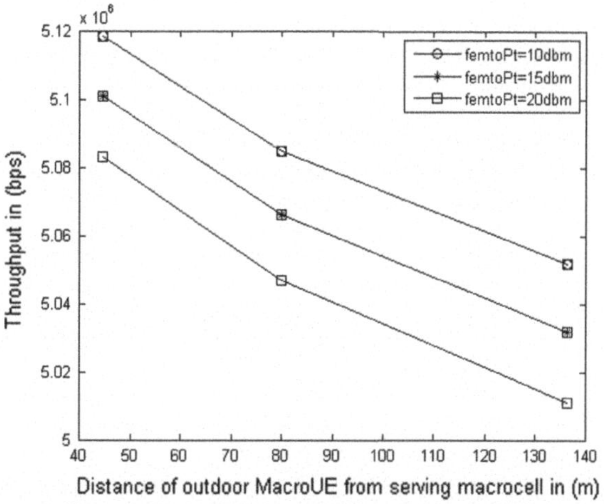

It is inferred that 2.8 and 3.2% of increased throughput are achieved when compared to indoor environment using 16-QAM and 64-QAM, respectively.

Hence, MUE (indoor and outdoor) experienced interference due to deployment; transmitting power of FBS is analyzed with respect to its received power from MBS and FBS. Based on received power, the enhancement in QoS of MUE is achieved.

5 Conclusion

The downlink interference problem in a co-channel deployment of FBS over MBS is analyzed. The performance analysis in terms of SINR and throughput of MUE is presented. The impact of MUE distance from MBS and deployment of FBS and their transmitting power are investigated. In future, the analysis can be extended to the channel model including shadowing and fading.

References

1. Lee YL, Chuah TC, Loo J, Vinel A (2014) IEEE journals and magazines on recent advances in radio resource management for heterogeneous LTE/LTE-A networks. IEEE Commun Surv Tutorials 16(4):2142–2180
2. Zahir T, Arshad K, Nakata A, Moessner (2013) IEEE journals and magazines on "Interference management in femtocells". IEEE Commun Surv Tutorials 15(1):293-311
3. Khan FH, Choi Y-J (2014) IET journals and magazines on "Adaptive mode configuration in two-tier macro-femtocell networks". IET Commun 8(7):1169–1179
4. Bouras C, Diles G, Kokkinos V, Kontodimas K, Papazois A (2014) Springer journal on" A simulation framework for evaluating interference mitigation techniques in heterogeneous cellular environments". Wirel Pers Commun 1213–1237
5. Chandhar P, Ghosh P, Das SS (2014) IEEE conference on "Performance analysis of Co-channel deployment of Femtocells with power control in 4G networks". In: Twentieth national conference on communications, 2014, pp 1–6
6. Kelif J-M, Diego W, Senecal S (2012) Impact of transmitting power on femto cells performance and coverage in heterogeneous wireless networks. In: IEEE conference on wireless communications and networking, 2012, pp 2996–3001
7. Siduo S, Lok TM (2013) IEEE journals and magazines on "Dynamic power allocation for downlink interference management in a two-tier OFDMA network". IEEE Trans Veh Technol 62(8):4120–4125
8. 3GPPTR 36.814 V9.0.0. (2010) Evolved universal terrestrial radio access (E-UTRA). Further advancements for E-UTRA physical layer aspects. (Release 9) in 3rd generation partnership project Technical report, pp 66–69
9. 3GPP TSG RAN WG4 (Radio) meeting, technical report. R4-092042: Simulation assumption and parameters for FDD HeNB RF Requirements. #51 San Francisco, CA, 4–8 May 2009

Mobile Foolproof Billing at Supermarkets

M. Abhiyukthana, C. Poovizhichelvi, P. Sindhuja, K. Srinivasan, B. Sharmila and R. Ramya

Abstract This paper briefs the design of a low cost, accurate, easy to use mobile billing in supermarket trolleys for minimizing the time wasted by customers standing in long queues for getting products billed and providing good customer satisfaction eliminating the need to compromise quality in interest of time-saving.

Keywords Low cost · Mobile billing · Supermarket · Trolley Time-saving · Customer satisfaction

1 Introduction

Nowadays, people in India prefer shopping at supermarkets and malls due to the availability of different varieties of products. It has actually become a kind of fashion trend in today's world. People find it interesting to buy things at a shopping mall. But a major drawback that leads to customer dissatisfaction is the time it takes for billing. People need to stand in long queues for getting their products billed [1]. In such situations, automatic billing at the time of inserting products into the trolley would be a welcoming idea. Many solutions pertaining to this idea have been proposed but have not yet been implemented. Hence, this paper has taken up an idea and designed a prototype for supermarket trolleys.

M. Abhiyukthana (✉) · C. Poovizhichelvi · P. Sindhuja · K. Srinivasan
B. Sharmila · R. Ramya
Department of Electronics and Instrumentation Engineering,
Sri Ramakrishna Engineering College, Coimbatore, Tamil Nadu, India
e-mail: abhiyukthana.1306002@srec.ac.in

B. Sharmila
e-mail: sharmila.rajesh@srec.ac.in

A. M. Zungeru et al. (eds.), *Wireless Communication Networks and Internet of Things*, Lecture Notes in Electrical Engineering 493,
https://doi.org/10.1007/978-981-10-8663-2_18

2 Literature Survey

Most of our markets are provided with the facility of barcode scanning of products. This is time-consuming process where the customer/buyer has to scan each and every product. In festive occasions, people who are standing in long queue will find difficult get over home. The shopkeepers also will find difficulty in managing the crowd. The barcode scanner has a sensor which converts vertical lines of different thicknesses into corresponding numerical data. For this, it needs a direct line of sight to the barcode for being read [2]. But it is not so in the case of an RFID-based system, the tags can be detected at greater distances (up to 300 ft). If a barcode is damaged, there are no ways to obtain the product details, whereas RFID tags are reusable and rugged due to the reason that they are covered by plastic coatings [3, 4]. This system is more advantageous than the existing methods. Figures 1 and 2 show the survey taken from two developed nations about the growth of supermarkets.

This statistics shows the percentage of people shopping in supermarkets. When the proposed system is implemented, it will certainly increase the growth of supermarkets and satisfaction of customers.

3 Proposed Method

Of the different methods proposed, the project has devised an easier system for sensing and processing of the billing amount. The information of the product is sensed, which contains passive RFID tag [3] on it, through an RFID reader. The information on the tag is used for adding the price of the product to the final billing amount. The final billing amount is transferred to the main computer via RF transmitter [5]. These programs are interfaced through an Arduino ATMEGA 328. Figure 3 shows the block diagram and the components used in the prototype model developed. The main concept deals with the objective of detecting the RFID tag and to extract the necessary information, i.e., the price of the product from it, add the price to the total billing amount. Figure 4 deals the flow process in the trolley. The following points describe the procedure to billing the products in the trolley.

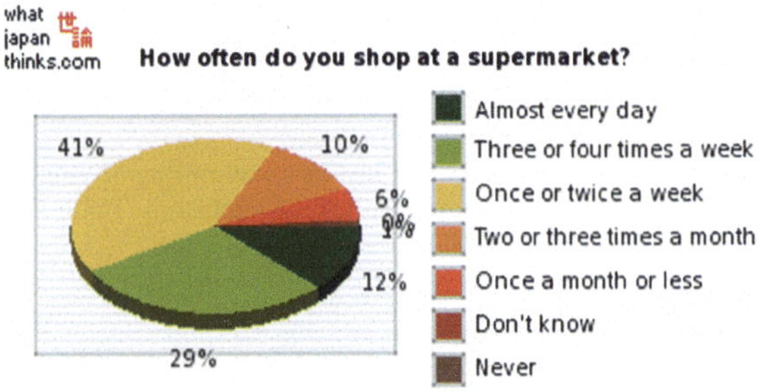

Fig. 1 Statistics showing the various frequencies of supermarket usage by people in Abroad (Japan)

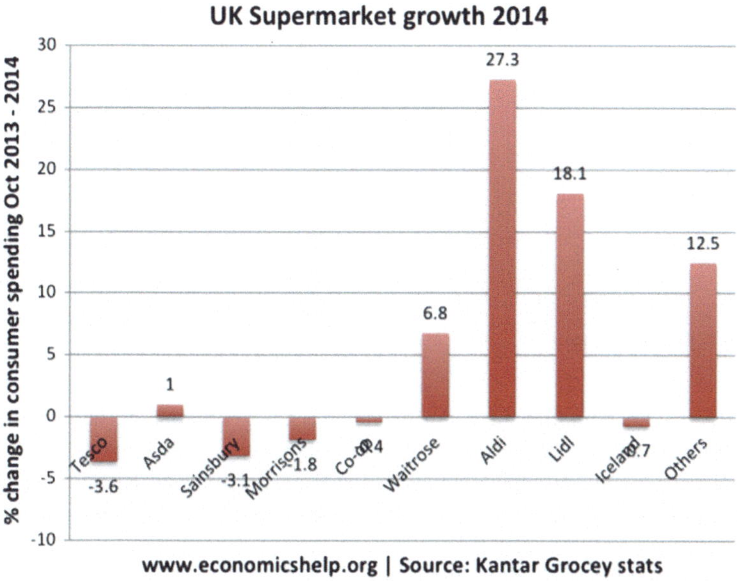

Fig. 2 Statistics showing the growth of supermarkets in UK in the year 2014

Fig. 3 An RFID tag

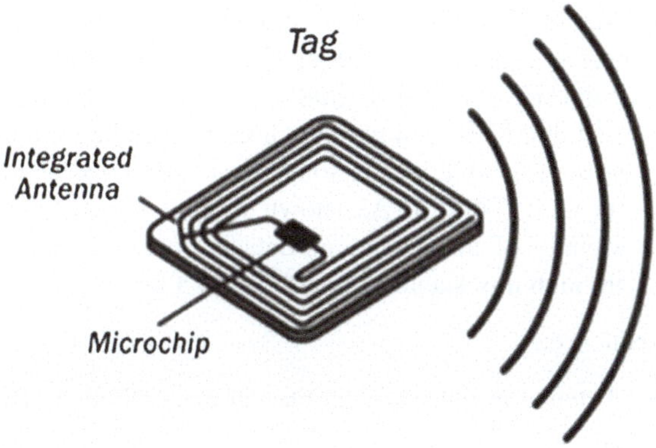

- Place the product with the RFID tag in front of the reader.
- The reader senses the tag.
- The ATMEGA 328 extracts the information on the tag and also counts the number of products in the trolley.
- The price of the product is fetched.
- The product's price is added to the main billing amount.
- If any product is removed from the trolley, the bill has to be deduced and also the count has to be reduced [4].
- Transfer the billing amount to the main computer when required.

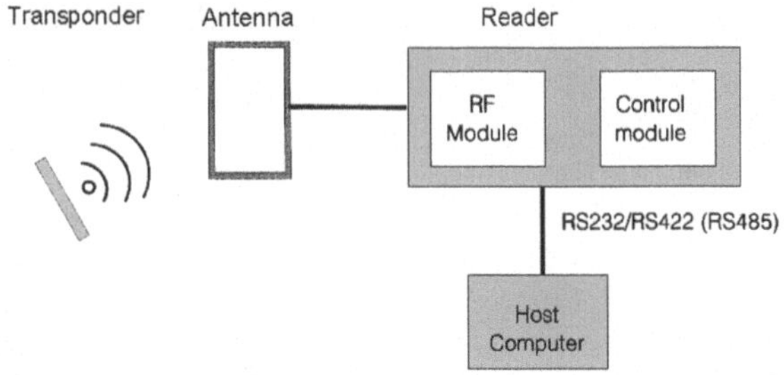

Fig. 4 RFID tag and reader representation

3.1 RFID System

An RFID system contains three components. These components have two parts each: a transceiver (transmitter/receiver) and an antenna forming an RFID reader; A transponder (transmitter/responder) and an antenna forming an RFID tag (Fig. 3). The RFID tag is read when the radio signal emitted by the reader strikes the tag and is reflected back to the reader [6].

There are two types of RFID tags:

- Passive RFID tags need an external energy source and rely on the energy given off by the reader. These are cheaper and hence are commonly used for consumer goods.
- Active RFID tags have their own power sources, which uses to generate a signal that is sensed by a reader. These are comparatively expensive, but can communicate over miles similar to radio wave communications. Hence, it can be said that RFID is a wireless link that can be used for uniquely identify an object or even people and is sometimes mentioned as dedicated short-range communication (DSRC). As shown in Fig. 4, when an RFID tag comes under the vicinity of the reader, its data is obtained by the reader and then transferred through any standard interfaces to a host [7].

Sequence:

- Antenna in the reader emits radio signals that activate the tag.

The emitted radio waves are in range from one inch to 100 ft or even more, depending upon the power output and frequency used. When the RFID tag interrupts the electromagnetic signals, it is detected by the reader.

- The reader then decodes the data that was encoded in the RFID tag's integrated circuit (silicon chip) after which the data is sent to the host computer for further processing.

This data transmitted is used for obtaining identification, location information, or in our case, specifics about the product tagged, such as price, weight, date of manufacturing.

3.2 Battery

The ARDUINO ATMEGA 328 needs an input power supply of 12 V and the RFID reader needs a power supply of 5 V; hence, a 12 V battery and a power supply module is used. Figure 5 shows how the various components are connected in the proposed circuit.

3.3 Arduino ATMEGA 328

The microcontroller used her is an Arduino UNO that is based on the ATMEGA 328. It has 40 pins in total, of which 14 are digital input/output pins (6 of which can be used as PWM outputs), 6 are analog inputs. In addition to these, there is a 16 MHz crystal oscillator, a USB connection, a power jack, an ICSP header, and a reset button. Figure 6 shows the pin configuration of an ATMEGA 328 chip.

3.4 RF Receiver and Transmitter

The final billing amount calculated has to be transferred to the main computer. As this communication is only one sided, we use an RF transmitter in the trolley and an RF receiver at the main computer.

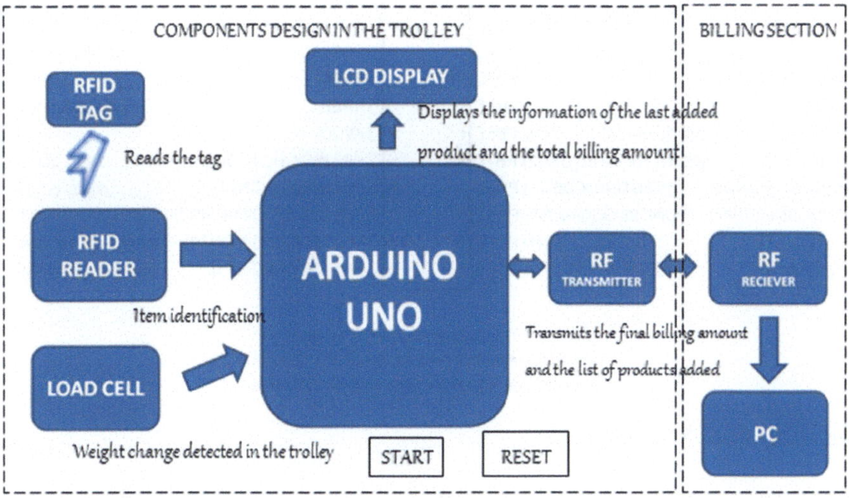

Fig. 5 Diagrammatic representation of the proposed supermarket trolley

The RF module used is an ASK hybrid transmitter + receiver one that uses a crystal-stabilized oscillator, that ensures an accurate frequency control for a good range of performance. There is no need for any external RF components except for an antenna. The module is ideal for remote control applications. The flow chart in Fig. 7 gives a better understanding of the steps involved in the billing process.

3.5 *Load Cell*

Since the system is to be made FOOLPROOF, load cell is used here for the prototype model. A 5 kg load cell is used.

3.6 *LCD*

The LCD used is ERM1604SYG-1 of 16 characters wide with 4 rows. The power supply range is 5 V, and the backlight used is yellow-green in color.

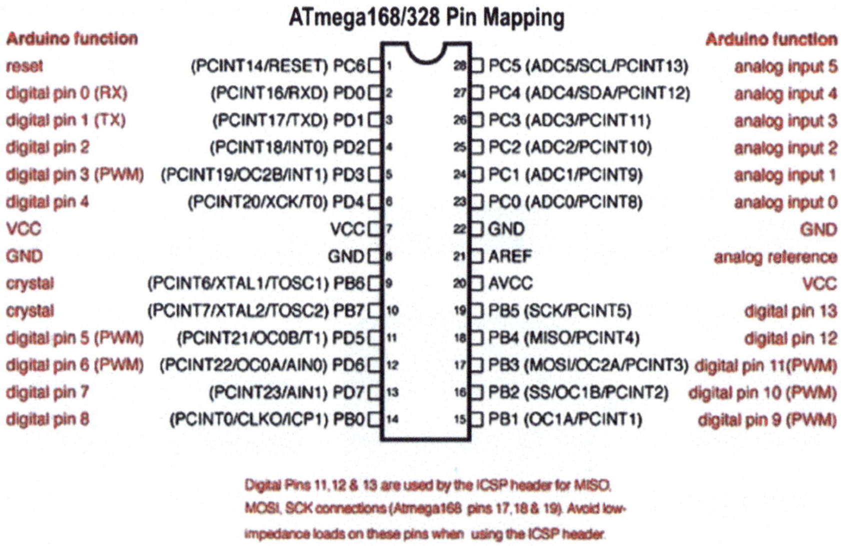

Fig. 6 Arduino UNO pin mapping

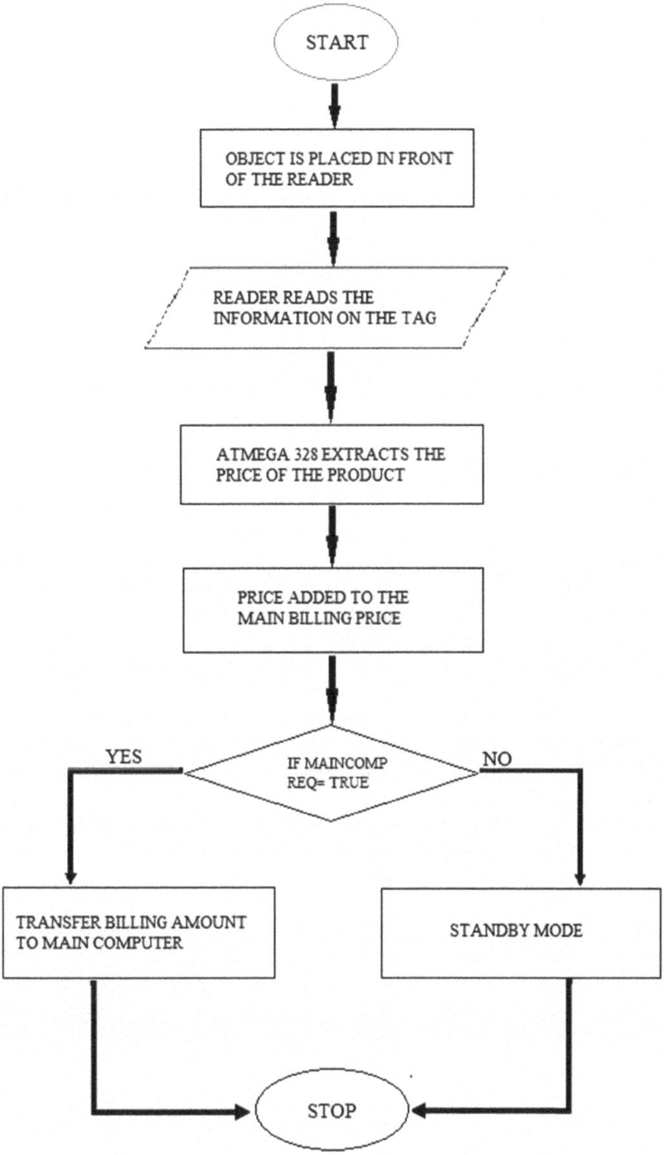

Fig. 7 Flowchart of the process in the trolley

4 Conclusion

The working product thus achieves the aim of facilitating automatic billing in shopping malls via shopping trolleys. This setup is handy and can be fixed in all carts in the supermarkets and other big shops. People would find this interesting and would make them happy during purchase. The time wasted in long queues for billing also is eliminated and thus produces a good shopping experience for the people.

Acknowledgements We thank the Management, Principal, Director Academics, Head of the Department, and Faculty Members of the Sri Ramakrishna Engineering College, Coimbatore, for their guidance and support. We thank our guide Mrs. R. Ramya, who gave us her valuable comments and shaped this project to perfection. We also thank our parents and friends who were kind enough to go through this project. We wish to thank our Institution and Alumni Cell to take this project to next level of development the actual product.

References

1. Hou J-L, Chen T-G (2011) An RFID-based shopping service system for retailers. Adv Eng Inform 25(1):103–115
2. Santos DSS, Pereira AMJ, Goncalves RMRM (2009) Intelligent Cart: architecture of an innovative system for the acquisition of products in grocery stores. Commun Int Bus Inf Manag Assoc J 8:80–87
3. Xiwen S (2012) Study on security issue of internet of things based on RFID. In: Proceedings of IEEE 2012 fourth international conference on computational and information sciences (ICCIS), Aug 2012, IEEE Press, pp 566–569. https://doi.org/10.1109/iccis.2012.301
4. Pagarkar M, Natesan M, Prakash B (2006) RFID in integrated order management systems, pp 6–10
5. Lekshmy S et al (2015) RFID based shopping trolley. Int J Comput Eng Res Trends 2 (12):1096–1099
6. Mohammadi S, Mesgarha SG (2011) Autonomous movement in car with the base of RFID. World Acad Sci Eng Technol 58:580–583
7. Gueaieb W, Miah MS (2008) An intelligent mobile robot navigation technique using RFID technology. IEEE Trans Instrum Meas 57(9)

Energy-Efficient-Based Optimizing Cluster Head Selection by Geometric-Based Mechanism and Implementation Using Soft Computing Techniques

S. Famila and A. Jawahar

Abstract The joint technique used is geometric-based analysis and transitional probability for the cluster head selection in wireless sensor networks with the combination of the threshold analysis and the trust factor for the determination of the optimized values by using soft computing techniques. In this paper, we present a geometric and the transitional probability for the cluster head selection in wireless sensor networks. Therefore, a continuous semi-Markov chain-based node behavior prediction process is incorporated for identifying the trust parameter that integrates the energy factor estimated previously based on fuzzy probability. The proposed algorithm is the fuzzy-based geometric translational probability (FGTP) for cluster head selection. The semi-Markov chain is implemented by using fuzzy logic for determining the cluster head selection process. The implementation of the proposed FGTCH scheme is investigated based on various performance metrics such as the Packet delivery ratio, Delay, Energy difference, Throughput and quantified trust. The trust parameter is incorporated in the proposed FGTCH scheme for optimizing the cluster head selection through soft computing techniques. To evaluate the proposed FGTCH model, extensive simulations were carried out in MATLAB. The number of nodes in a network ranges from 70 to 100 nodes. The simulation is conducted to evaluate the performance of the proposed FGTCH, and it is compared with existing LEACH for clusters head formation, the average end-to-end delay, packet delivery ratio, and lifetime computation. When the number of nodes is 30, the percentage of average end-to-end delay (s) of cluster formation using fuzzy-based geometric translational probability for cluster head selection algorithm is decreased by 4.87% than LEACH. When the number of nodes is 30, the percentage of packet delivery ratio of cluster head selection using fuzzy-based geometric translational probability algorithm is increased by 6.60% than LEACH.

S. Famila (✉) · A. Jawahar
Department of ECE, SSN Engineering College, Kalavakkam, Chennai, India
e-mail: selvafam@rediffmail.com

A. Jawahar
e-mail: jawahara2@ssn.edu.in

A. M. Zungeru et al. (eds.), *Wireless Communication Networks and Internet of Things*, Lecture Notes in Electrical Engineering 493,
https://doi.org/10.1007/978-981-10-8663-2_19

When the number of nodes is 80, the percentage of packet delivery ratio of cluster formation using fuzzy-based geometric translational probability algorithm is decreased by 10.42% than LEACH and the existing methods.

Keywords FGTCH · Average end-to-end delay · Packet delivery ratio Throughput · Semi-Markov chain · LEACH

1 Introduction

A wireless sensor network (WSN) is a cooperative network in which collection of nodes is organized and works in a coordinated way. Each and every node has individual processing capability by means of one or more microcontrollers and CPU chips, memory elements, radio frequency transceiver, power source, and different types of sensor and actuators [1]. It is a self-organized network in which nodes are deployed in an ad hoc manner and communicate with each other wirelessly. From the recent past, the WSN enabled with recent technological developments in wireless communication finds applications in battlefield surveillance and any other critical environment monitoring [2]. The major challenges of this wireless sensor networks are: (i) the wireless sensor nodes are small in size, consume little amount of energy, restore limited resources. (ii) The energy of the nodes cannot be refilled or reloaded. (iii) The nodes remain autonomous and operate in an unattended fashion which makes the topology of the network mysterious.

Since the wireless sensor environment is highly resource-constrained, the most important thing for sensor nodes is to preserve energy resources in order to sustain the performance of network. Initially, the research works in WSN focus toward designing an optimized communication standard, but those works do not concentrate on optimizing the usage of sensor nodes' energy and power [3]. The recent works on sensor networks address the basic problems of cluster head election in order to efficiently deploy the wireless sensor network with optimized cost and quality [4]. The problems addressed by recent literature are clustering of sensor nodes for cooperative network, election of an optimized node as a leader or cluster head for effective communication and energy indulgence. There are also many other literature that focus on the clustering algorithms for grouping the sensor nodes for efficient energy management within the nodes.

The most important problem addressed in the proposed FGTCH scheme focuses on the need for facilitating significant solution for the factors that influences the formation of cluster in order to ensure superior performance of the network. In the traditional clustering techniques, the cluster heads are initially selected at the time of network implementation i.e. at the static time. The characteristics used for static time cluster head selection are the size, area and members involved in the cluster. The standard methodology of cluster head selection is generally based on the sensor node that possesses adequate battery power and high signal-to-noise ratio [5]. However, electing a cluster leader in such way will be a complex process. Recently, the heuristic approaches such as fuzzy logic (FL) and genetic algorithms (GA) have proved to be efficient for various applications in wireless networks. In this paper, the selection of new cluster head is incorporated using the concept of the fuzzy logic. The proposed fuzzy logic-based cluster head selection is evaluated through simulation experiments.

The remaining sections of the paper are organized as follows. Section 2 presents a short view on some of the related works that are contributed to fuzzy-based cluster head selection in the literature. The phase-by-phase process of FGTCHS mechanism is detailed in Sect. 3. Section 4 highlights the inference of results and discussions that portray the performance of the proposed FGTCHS. Section 5 depicts the major contribution of FGTCHS approach with the scope of their future plan.

2 Related Work

From the recent decade, a number of fuzzy-based cluster head election schemes have been proposed for sensor environments. Some of the significant approaches are highlighted with their merits and demerits.

Anno et al. [6] proposed an efficient way of selecting a cluster head selection that plays a vital role in the energy management of the wireless sensor networks. The selection of cluster head by fuzzy-based system in sensor networks has overcome the challenges and increased the energy efficiency. The two fuzzy-based systems for cluster head selection are the FCHS System 1 and FCHS System 2. The proposed FCHS System is implemented based on the chosen parameters such as distance of cluster centroid (DS), remaining battery power of sensor (SP) and network traffic (NT). The FCHS System 2 has chosen the various parameters such as remaining battery power of sensor (RPS), degree of number of neighbor nodes (D3N), distance from cluster centroid (DCC). The FCHS System 2 shows improved performance than the existing system and FCHS System 1. The remained energy of the sensor and the number of neighbor nodes are more important parameters for CH selection than the distance of the node from the sink. Gajjar [7] proposed that the WSN has a unique parameter in which it senses the data and communicates in wireless and then communicates it to a base station. Cluster head is selected by fuzzy logic, and the protocol is CHUFL. The parameter used for the CHs selection is the residual energy, reachability from its neighboring node, and distance from base station as fuzzy input variables for cluster head selection. The comparative study of cluster head selection in various works by Kim et al. and cluster head selection method for wireless sensor networks based on fuzzy logic by Anno et al. show that CHUFL is up to 20% more energy efficient and sends 72% more packets to base station compared to AODV protocol. Natarajan [8] contributed a cluster head selection which is based on the new approach as the input parameter depends on the entire communication, and a high energy node has to be selected as cluster head. A novel predictive fuzzy-based cluster head selection algorithm is proposed. This proposal suggests a new input parameter called the amount of recurrent

Communication apart from the standard parameters such as the residual Power of Sensor Nodes, Degree of Neighboring Nodes, Distance between the Node and Base Station and the Sensor Node Movement that is potential for the cluster head selection. Kiran [9] proposed a cluster head with the basis of the three parameters such as the residual energy of node, the vicinity, and the packet delivery ratio with respect to distance from the BSs, and all the parameters are incorporated, and it is selected based on the fuzzy logic and the result in which the energy consumption is more efficient than the previous fuzzy-based logic and proves a long lifetime network.

3 Proposed Work

The proposed algorithm relies on (FGTP) for cluster head selection that inspires semi-Markov chain using *t*-value. This FGTP plays a significant role in initializing the number of sensors for exhibiting the action of cluster heads in each round. The optimal FGTP-value estimated through semi-Markov process is found to depend on cluster size and density of the sensors deployed. Thus, the determination of optimal FGTP-value has to be carefully determined for improving the quality of cluster head election which in turn reduces the message overhead to a significant level. In case of high-density sensor network, the optimal FGTP-value systematically decreases as the number of capable cluster head of the network increases. However, higher FGTP-value will also impact and influence the network performance as it utilizes maximum amount of energy under cluster head election competition. This cluster head competition not only bounds to initiate more number of cluster heads but also reduces the enabled cluster size compared to the deployed size of the cluster. In contrary, lower FGTP-value on the other hand increases the enabled cluster size which is comparatively greater than the deployed cluster size. In the baseline approaches like UCR and EC algorithms, the deployed sensor nodes are forced to use the same FGTP-value which was preset as 0.2 and 0.1, respectively. But in the proposed FGTPCHS, the value of FGTP is dynamically updated, and they are unique in each of the sensors deployed. The potential with which a sensor node dynamically updates its FGTP-value for initializing itself as the cluster head purely depends on its estimated initial FGTP-value in each round. Initially, each and every sensor is set to possess the same value of FGTP, but when time progresses, the FGTP-value gets dynamically updated. This FGTP-value is systematically updated in each round so as to identify an operated cluster size close to the deployed size of cluster. Then, the sensors start its process in the formation of cluster, collecting data and aggregating them to the sink sensor node.

The process of cluster formation is divided into four phases such as (i) co-coordinative cluster head election based on FGTP, (ii) cluster head election-based competition, (iii) discovery phase for identifying clusters, (iv) association phase of clusters, and (v) confirmation phase of cluster head election. In phase 1, the value of FGTP is estimated through a semi-Markov-based weighted function which is predominantly used for cluster formation in order to adjust the present value of initial FGTP that needs to be updated in the next subsequent rounds. In phase 1, the act of election is achieved by the sensors elected through the initiated FTGP value. Then, based on the packet forwarding potential and energy availability of the sensor nodes, each sensor computes a normalized probabilistic value called 'P_{deter}' that ranges between 0 and 1. If P_{deter} is found to be less than or equal to the value of FTGP, each sensor node identifies itself as the co-coordinative cluster head in a distributive manner and then calculates delay based on the inverse function of available residual energy possessed by a specific sensor node. When the remaining sensor nodes get withdrawn from competition and transit to seeping state, the start of the cluster discovery process is performed until when the value of P_{deter} determined is found to be greater than the value of FTGP.

In phase 2, only the eligible co-coordinative elected cluster heads enter for competition. Each eligible cluster head sends the cluster head advertisement to the closest proximity located sensor nodes of distance 'D_{eligible}' only when the time of expiry in delay as estimated in the cluster head election phase is confirmed. If the cluster head advertisement is not received by the eligible cluster heads, then they enter into sleep mode and they are assumed to wake up only during the next phase of cluster discovery phase. In phase 3, the cluster discovery process is initiated by the cluster heads through their own advertisement using cluster head discovery packet. The sensor nodes within the distance 'D_{eligible}' are considered as cluster members. Each and every cluster member listens to the cluster head discovery packets and then chooses closest cluster head if more than one cluster are situated in the closest proximity. Further, a significant parameter called extended cluster head advertisement radius (ECHAR) is used for facilitating 99% confidential limits such that each eligible cluster will be associated to a minimum cluster member under communication. After the assurance of this connectivity, cluster member as well as cluster heads focuses on cluster association and confirmation phase.

In the final phase, cluster members send the cluster association packets to their associated cluster heads that contain the available residual energy of the advertising cluster members. Then, the cluster heads respond with the aid of the cluster head acknowledgment packets. In addition, the cluster heads compute the cluster size of operation by determining the proximal distance between each cluster head and their associated cluster member of farthest distance when the entire association packets reach them with the delay period. The cluster heads estimate the mean residual energy of the cluster based on the information about residual energy received from the cluster members. Then, each cluster head needs to allocate time period for their associated cluster members using a specific kind of packet called time period confirming packet. This time period confirming packet also elucidates two additional information pertaining to estimated cluster size under operation and mean available cluster energy of the sensor nodes. Hence, the cluster members receive the estimated assignment of time slot for updating time radios and variable residual energy for fuzzy classification using semi-Markov process for adjusting t-value-inspired FGTP-value for using it in the next round of operation.

4 Simulation Experiments and Discussions

The performance of FGTCHS is evaluated using NS-2.33 based on energy consumptions and packet delivery ratio.

Figure 1 represents the performance of FGTCHS based on energy consumptions. The energy consumptions of FGTCHS are found to phenomenally improve when compared to LEACH and LUCA. This systematic improvement of energy consumptions is mainly due to the incorporation of *t*-value that statistically investigates the election of cluster head through the computation of mean and standard deviation from the elucidated data that effectively quantifies the election of cluster head. Thus, FGTCHS is found to improve the packet delivery ratio by 15–18% than LEACH and 22–25% than LUCA.

Figure 2 represents the performance of FGTCHS based on packet delivery rate. The packet delivery rate of FGTCHS increases systematically when compared to LEACH and LUCA. This systematic improvement in packet delivery rate is due to the use of fuzzy rules that categorizes malicious behavior from cooperative behavior. This categorization of behavior facilitates rapid selection of cluster head

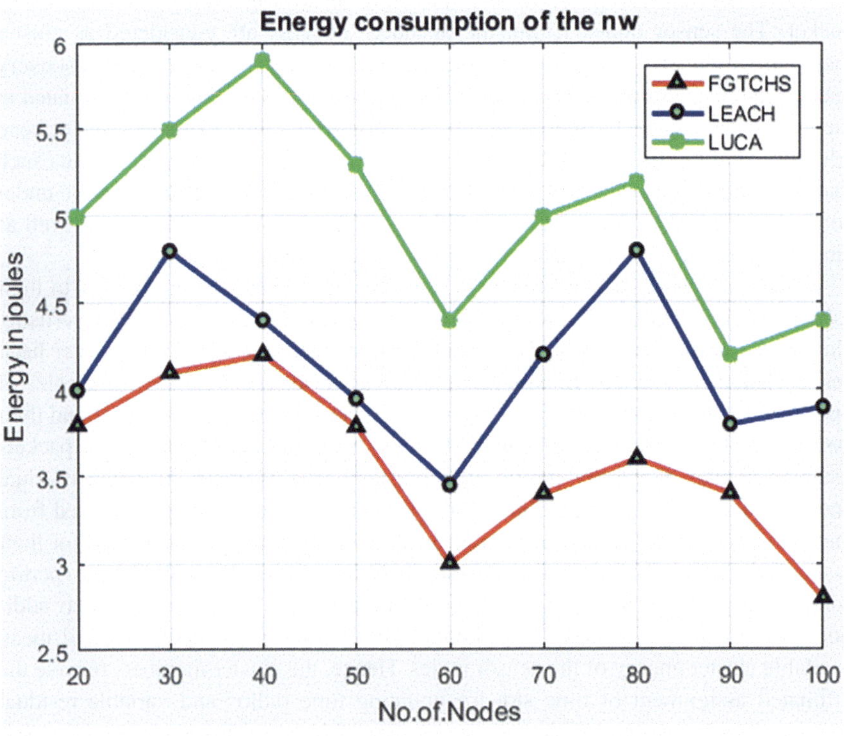

Fig. 1 Performance of FGTCHS—energy consumptions

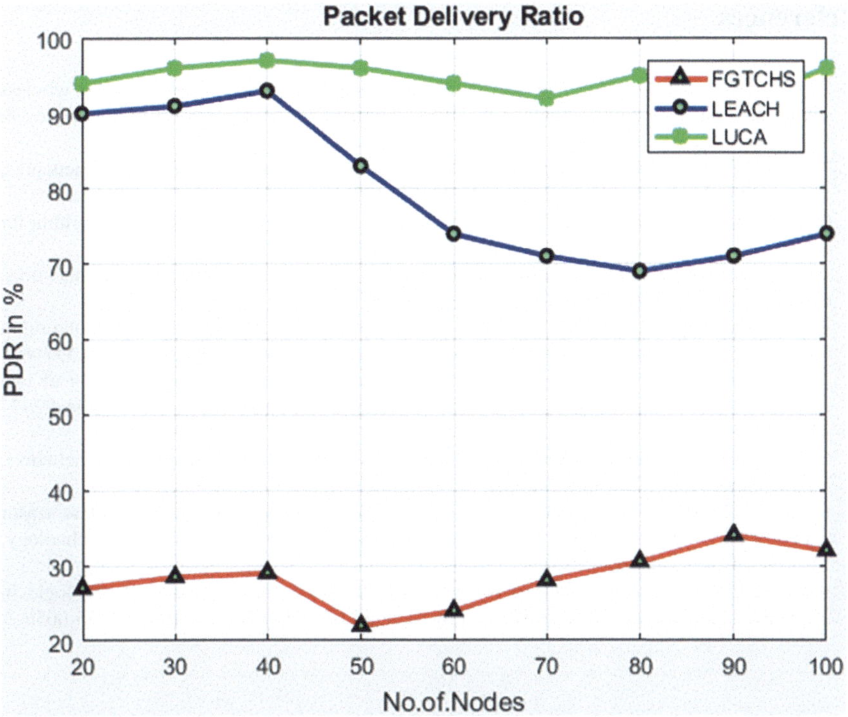

Fig. 2 Performance of FGTCHS—packet delivery ratio

and prevents selfish cluster head from the network for improving the lifetime of the network. Thus, FGTCHS is found to improve the packet delivery ratio by 14–19% than LEACH and 17–27% than LUCA. In addition, on an average, the performance of FGTCHS is found to improve by 16% in terms of packet delivery ratio and reduces the energy consumptions by 18% than LUCA.

5 Conclusion

In this paper, FGTCHS is proposed for facilitating reliable cluster head election in order to improve the lifetime of the network. This FGTP produces a positive impact on the network since it conserves maximum amount of energy during the process of cluster head election competition. The simulation results of FGTCHS clearly prove that on an average, the performance of FGTCHS is found to improve the packet delivery rate and throughput to a minimum range of 12% and maximum of 165% in the superior performance in contrast to LEACH and LUCA. In the near future, it is also planned to formulate a fuzzy-inspired cluster head election that relies on meta-heuristic optimization techniques for effective elections of noble cluster heads in order to improve the lifetime of the network.

References

1. Li C, Ye M, Chen G (2005) An energy-efficient unequal clustering mechanism for wireless sensor networks. In: Proceedings of the IEEE international conference on mobile adhoc & sensor systems, pp 597–604
2. Tang J, Wang Y (2013) Improved EEUC routing protocol for wireless sensor networks. J Chongqing Univ Posts Telecommun 25(2):172–177
3. Qing Z, Chai Q, Liu L (2008) Unequal scaled energy-efficient clustering routing algorithm for wireless sensor networks. Comput Eng 34(23):98–100
4. Abbasi AA, Younis M (2007) A survey on clustering algorithms for wireless sensor networks. Comput Commun 30(2):2826–2841
5. Rizvi S, Qureshi HK, Khayam SA (2012) A1: an energy efficient topology control algorithm for connected area coverage in wireless sensor networks. J Netw Comput Appl 35(2):597–605
6. Anno J, Barolli L, Durresi A, Xhafa F, Koyama A (2008) Performance evaluation of two fuzzy-based cluster head selection systems for wireless sensor networks. Mob Inf Syst 4:297–312 (IOS Press)
7. Gajjar S (2014) Cluster head selection protocol using fuzzy logic for wireless sensor networks. Int J Comput Appl 97(7) (0975–8887)
8. Natarajan H (2014) Fuzzy based predictive cluster head selection scheme for wireless sensor networks. In: Proceeding of the 8th international conference on sensing technology, Liverpool U.K., 2–4 Sept 2014
9. Kiran J (2016) An augmented approach for cluster head selection using fuzzy logic in clustering hierarchy for WSN. Int Res J Eng Technol (IRJET) 03(10). e-ISSN: 2395-0056

Enhancement of QOS Parameters in Cluster-Based Wireless Sensor Network Using Cooperative MIMO

R. Guhan, U. Hari and B. Ramachandran

Abstract Sensor nodes are deployed in the wireless sensor network with fixed energy and positioned randomly. For sustaining the power various multihop routing protocols are proposed. Cooperative MIMO scheme is one of the techniques where more nodes present near to the sink suffers traffic burden to forward the data from other cluster head in the network that causes hot spot problem. In order to overcome this problem we propose mobile sink technique in cooperative MIMO based on unequal clustering. In this technique, the mobile sink are made to move back and forth along the boundary near to the base station which collects the data in its vicinity, forwards it to the base station. An alternate relay node selection scheme is also accomplished for further enhancement. Simulation results show that proposed work improves the network lifetime over unequal clustering multihop routing protocol (UCMR) and cooperative MIMO scheme.

Keywords Mobile sink · Cooperative MIMO · Unequal clustering
Relay nodes

1 Introduction

In wireless sensor network large numbers of sensor nodes are distributed randomly in the sensor field to sense the environment and to impart the recognized data to the base station [1, 2]. The wide application of WSN includes environment sensing, health care monitoring, industrial monitoring, and natural disaster prevention, etc.

R. Guhan (✉) · U. Hari · B. Ramachandran
Department of Electronics and Communication Engineering, SRM University, Kattankulathur, Chennai, India
e-mail: guhan_r@srmuniv.edu.in

U. Hari
e-mail: hari.u@ktr.srmuniv.ac.in

B. Ramachandran
e-mail: ramachandran.b@ktr.srmuniv.edu.in

A. M. Zungeru et al. (eds.), *Wireless Communication Networks and Internet of Things*, Lecture Notes in Electrical Engineering 493,
https://doi.org/10.1007/978-981-10-8663-2_20

[1, 3]. Sensing unit, computing unit, and communication unit are the three components of the sensor node. Computing unit process the collected raw data and communication unit comprises of transmitter and receiver. With limited power, extending energy efficiency and network lifetime become an important goal.

In order to sustain the power, cooperative MIMO concepts are implemented in which group of sensor nodes act as the virtual antenna for providing reliable communication that is based on multihop routing scheme leads to hot spot problem [4]. To overcome this, cooperative MIMO based on unequal clustering proposed [5, 6] and to further elevate the performance, mobile sink technique is implemented in which the sink moves in a predetermined path that gathers the packets and delivers it to the base station [7].

The remaining part of the paper is organized as follows. Related work is discussed in Sect. 2. The proposed work is discussed in Sect. 3. The energy consumption model is proposed in Sect. 4. Simulation results are discussed in Sect. 5, and conclusion is drawn in Sect. 6.

2 Related Work

In wireless sensor network, various algorithms are put forward to extend the network lifetime. LEACH [1–3] is self-organized clustering algorithm based on single hop communication where the nodes are organized into specific networks and the member nodes present within the network that transmits the data to the cluster head in their schedule. HEED [1, 3] extends the LEACH, cluster heads are elected here based on the residual energy, and communication cost results in the distribution of energy consumption. The drawbacks of the single hop communication are the nodes farthest from base station exhaust quickly, while in multihop technique nodes near to based station are overloaded with more traffic results in hot spot problem.

To circumvent the above problem in [5, 6], unequal clustering multihop routing protocol UCMR is suggested in which the network size is arranged with respect to base station, that is to say, clusters size nearby to the base station is small in size and increases with respect to the base station.

MIMO concepts are executed in WSN to reduce the consumption of energy; but possessing multiple antennas leads to energy draining mechanisms in nodes hence, sharing antennas by cooperative nodes leads to achieve spatial diversity. Multiple sensor nodes are cooperatively used for trans-receiving operation [4, 8].

In [8], C-LEACH scheme cooperative MIMO techniques are incorporated in LEACH to increase lifetime; here, the cluster head elects the cooperative nodes within the cluster and transmits the data effectively. The disadvantage of this scheme is when there is imperfect data aggregation, the amount of data that each cluster head transmits varies widely results in uneven distribution of energy.

In [4], CH-C-LEACH scheme is proposed to overcome the above limitations and to further enhance the lifetime. Here the cluster head gets coupled with the other cluster head that transfers the data cooperatively to the base station results in even disperse of energy and the load is balanced more beneficial.

To enhance the performance in this paper, CH-C-LEACH is implemented in unequal clustering. However when a sensor network implemented with a fixed sink results in hot spot problem. To solve this in [7, 9] mobile sink technique is proposed that it has a pre-decided path for mobile sink to move which is situated in the center of the field.

Cluster formations are done here based on stable election protocol SEP [10] where the cluster gathers the packets and delivers it to the mobile sink. To further enhance the performance in this paper, higher energy nodes present near to the cluster head are chosen as the alternative relay node, and mobile sink technique is implemented in the vertical boundary of the sensor field.

3 Proposed Scheme

Consider a network with $M \times M$ square meter and N sensor nodes deployed randomly, and nodes are assumed to be stationary having same initial energy, unique ID, and sink node which are not static. The nodes are sectionally organized into clusters containing member nodes, cluster head, and relay nodes. The proposed model is illustrated in Fig. 1.

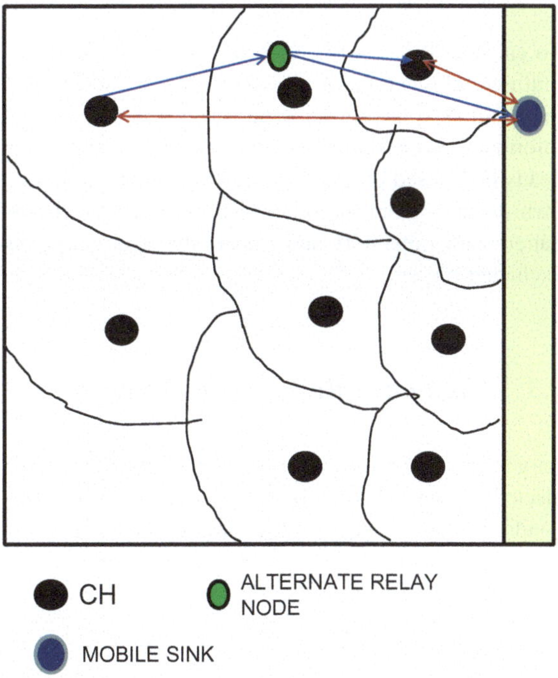

Fig. 1 Mobile sink model

3.1 Cluster Formation Phase

Initially, the sensor field is partitioned into sectors. Nodes present in the center of each sector having more surrounding nodes nearer to the center of the cluster are chosen as the cluster head.

In [5] the successive rounds, the current cluster head elects the new cluster head by evaluating cost factor of the nodes in each cluster

$$F_i = \frac{D_i}{1/M.\sum_{K=1}^{M}(D_k)} \frac{1/M.\sum_{K=1}^{M}(D_K)}{C_i} \frac{E_{ri}}{E_O} \tag{1}$$

where E_{ri} is the residual energy of the node i, D_i denotes the degree of the node i, C_i nodes distance from the centroid of the cluster, E_0 denotes the initial energy of the node i.

Elected cluster head broadcasts the ADV messages to its member nodes, upon receiving it, the member nodes transmit the Join – REQ back to the cluster head. Since UCMR concept is implemented the radius of the clusters are varied with respect to the distance from the base station. That is to say, smaller size clusters exist near to the base station.

3.2 Cooperative Node Selection

To enhance the network lifetime the cluster head are chosen as cooperative transmitting and receiving nodes. The member nodes of the clusters sends their sensed information to the cluster head and it performs the data aggregation, distributes the information containing its own volume of data, that are registered other cluster heads [6]. Based on the broadcasting signal strength, the cluster head determine its own distance from the other CH and once the distance is known the CH will get paired with other CH and impart the data cooperatively to the sink during data exchange phase.

3.3 Alternative Relay Node Selection

Larger energy nodes present in the center of the field near to the cluster head are elected as the substitutive relay node for redirecting the data to sink under the condition if the energy of the current cluster head diminishes.

3.4 Mobile Sink Implementation

Higher energy nodes near to the cluster head are regarded as the relay nodes. When one or more relay nodes present near to the cluster head then the nodes are chosen based on compass routing scheme by which nodes are elected based on minimum angle between the line-joining cluster head with mobile sink and the higher energy nodes [2]. Data from the chosen relay nodes are forwarded to the mobile sink projected in the perpendicular right end of the field.

4 Energy Consumption Analysis

The energy consumption of the whole network can be computed by modeling the energy consumption analysis of one bit. The average total power consumption comprises of two components, consumption of power by the power amplifiers and the circuit blocks.

Energy consumption in transmitting one bit is given by [11]

$$E_{\mathrm{bt}} = (P_{\mathrm{PA}} + P_{\mathrm{C}})/R_{\mathrm{b}} \tag{2}$$

$$P_{\mathrm{PA}} = (1+\alpha)P_{\mathrm{out}} \tag{3}$$

Power utilization of the power amplifier is roughly given above, where P_{C} is the power consumed by the circuit blocks, R_{b} is the system bit rate and α is the radio frequency power amplifier's efficiency

$$P_{out} = E_{\mathrm{b}}R_{\mathrm{b}}\frac{(4\pi)^2 M_{\mathrm{l}}\Psi_{\mathrm{f}}}{\lambda^2 G_{\mathrm{t}}G_{\mathrm{r}}}d^2, \quad d \le d_{\mathrm{o}} \tag{4}$$

$$P_{out} = E_{\mathrm{b}}R_{\mathrm{b}}\frac{M_{\mathrm{l}}N_{\mathrm{f}}}{G_{\mathrm{t}}G_{\mathrm{r}}h_{\mathrm{t}}^2 h_{\mathrm{r}}^2}d^4, \quad d \ge d_{\mathrm{o}} \tag{5}$$

where E_{b} denotes the energy per bit for given BER at the receiver, R_{b} indicates the system bit error rate, d denotes the transmission distance, distance threshold is given by d_{o}, G_{t} and G_{r} are the transmitter and receiver antenna gain, h_{t} and h_{r} are the height of the transmitting and receiving antenna, noise figure of the receiver ids given by $\Psi_{\mathrm{f}} = \frac{\Psi_{\mathrm{r}}}{\Psi_{\mathrm{o}}}$, where Ψ_{r} is the power spectral density of the effective noise at input of the receiver, Ψ_{o} is the power spectral density measured at room temperature.

Power consumption of transmitting and receiving circuit blocks can be approximated by

$$P_{\mathrm{ct}} = P_{\mathrm{DAC}} + P_{\mathrm{elec}} \tag{6}$$

$$P_{\text{cr}} = P_{\text{LNA}} + P_{\text{IFA}} + P_{\text{ADC}} + P_{\text{elec}} \tag{7}$$

where

P_{elec} denotes the power utilization of frequency synthesizer, active filter component, and mixer.

P_{LNA} denotes the power dissipation by the low noise amplifier.

P_{IFA} denotes the consumption of power by the intermediate frequency amplifier.

P_{ADC}, P_{DAC} represents the power utilization of analog to digital converter and digital to analog converter.

The total circuit power can be obtained by adding the above two equations

$$P_{\text{c}} = N_{\text{t}}P_{\text{ct}} + N_{\text{r}}P_{\text{cr}} \tag{8}$$

where N_{t}, N_{r} are the number of transmitting and receiving circuits.

Energy consumption per bit for cooperative MIMO implemented in unequal clustering is given by

$$E_{bit} = (1+\alpha)\rho N_{\text{o}} \frac{(4\pi d)^2 M_{\text{l}} \Psi_{\text{f}}}{\lambda^2 G_{\text{t}} G_{\text{r}}} + \frac{P_{\text{tx}}}{R_{\text{b}}} n_{\text{T}} + \frac{P_{\text{rx}}}{R_{\text{b}}} n_{\text{R}} \tag{9}$$

where AWGN power spectral density is given by N_{o}, P_{tx}, and P_{rx} denotes the power dissipation at transmitter and receiver circuits, R_{b} represents the transmission rate, α depends on power amplifiers drain efficiency, ρ denotes SNR at the receiver, distance between transmitter and receiver is denoted by d, power utilization values are, P_{tx} = 38 mV, P_{rx} = 41 mV of TelosB mote.

5 Simulation Results

The analysis of the proposed scheme is carried out using MATLAB. Sensing field of dimension 250 * 250 is considered. 200 static nodes are deployed randomly in the field. The initial energy of the nodes is set to 1 J, and mobile sink is assumed to be located at three instants (230, 180), (230, 100), (230, 40) in the right-hand side vertical boundary of the sensor field. The simulation parameters are listed in Table 1.

Table 1 Simulation parameters

Simulation parameter	Value
Network coverage	250 m × 200 m
Bs location	230, 100 m
No. of nodes (N)	200
Efficiency of RF power amplifier (α)	0.4706

(continued)

Table 1 (continued)

Simulation parameter	Value
Link margin (Ml)	40 dB
AWGN power spectral density (N_o)	−171 dB/Hz
Receiver noise figure (N_f)	10 dB
Carrier frequency (f_c)	2.5 GHz
Bandwidth (B)	20 kHz
Transmission rate (R_b)	250 kbps
Circuit power consumption of transmitter (P_{ct})	98.2 mW
Circuit power consumption of receiver (P_{cr})	112.6 mW
Antenna gain of transmitter and receiver (G_t, G_r)	5 dB

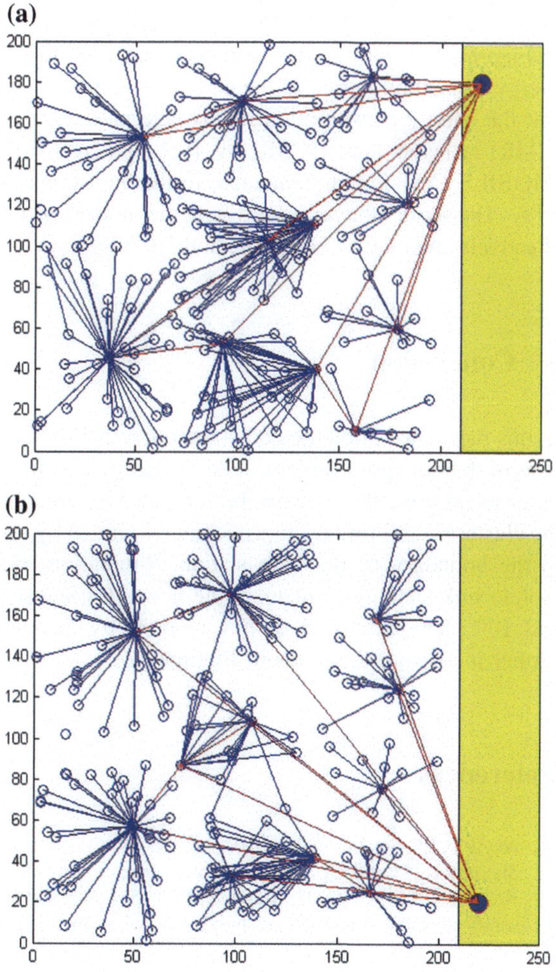

Fig. 2 **a**, **b** Network topology with mobile sink at various positions

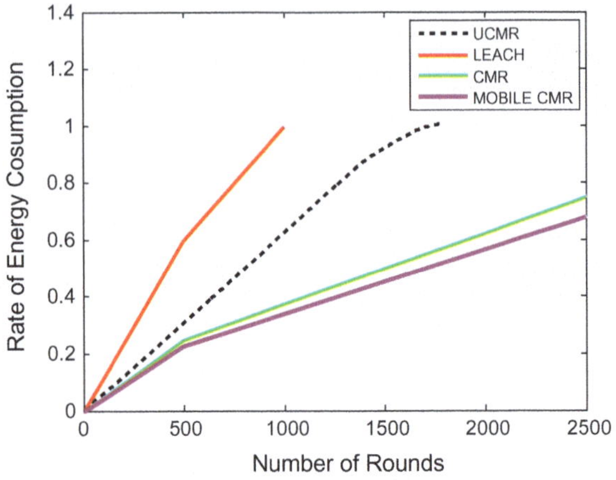

Fig. 3 Energy consumption rate of the nodes

Figure 2a, b shows the simulated network topology. Figure 3 illustrates the energy consumption rate of the nodes for various rounds. It is observed from Fig. 3 that the energy consumption rate of alternate node enacted in cooperative MIMO (CMR) is lesser than UCMR by 10% and by LEACH 15%. When mobile sink (MOBILE CMR) is implemented, further conservation of energy is achieved up to 20%. Therefore, the proposed scheme improves the energy conservation more effectively than LEACH, CMR, and UCMR.

6 Conclusion

In this paper, we implemented cooperative MIMO scheme in unequal clustering to control the hot spot problem. Rate of energy consumption is reduced effectively in order to improve the network performance by electing alternate relay node near to the cluster head. Further, improvement is shown by accomplishing mobile sink near to the boundary of the base station. Simulation results prove that the proposed mobile sink technique enhances the network lifetime over UCMR and CMR by 20 and 10%. The extension of this work is by incooperating genetic algorithm for further increasing the network lifetime.

References

1. Akyildiz IF, can Varun M (2010) Wireless sensor networks. Wiley
2. Shoraby K, Minoli D, Znati T. Wireless sensor networks, pp 222–224
3. Kazerooni AA, Jelodar H, Aramideh J (2015) Leach and heed clustering algorithm in wireless sensor network: a qualitative study. Adv Sci Tech 9(25):7–11

4. Vidhya J, Danajayan P (2010) Energy efficient STBC-encoded cooperative MIMO routing scheme for cluster based WSN. Int J Commun Netw Inf Secur (IJCNIS) 2(3)
5. Hari U, Ramachandran B, Johnson C (2013) An unequally clustered multihop routing protocol for WSN. In: International conference on advances computing, communications and informatics (ICACCI) 2013, pp 1007–1011
6. Yuan H, Liu Y, Yu J (2011) A new energy efficient unequal clustering algorithm for WSN. IEEE
7. Wang J, Zhang Z, Xia F, Yuan W (2013) An energy efficient stable election-based routing algorithm for WSN. MDPI, 24 Oct 2013
8. Vidhya J, Danajayan P (2010) Lifetime maximization of multihop WSN using cluster based cooperative MIMO scheme. Int J Comput Theor Eng 2(1)
9. Heidari M, Noorinamzade A, Reza Naji H (2015) Effect of using mobile sink and clustering on energy consumption in WSN. Turk J Electr Eng Comport Sci
10. Smaragdakis G, Matta I, Bestavros A (2004) SEP: a stable election protocol for clustered heterogeneous wireless sensor networks. In: Proceedings of the international workshop on sensor and actor network protocols and applications (SANPA 2004), Boston, MT, USA, 22 Aug 2004
11. Behera TM, Singh SS (2012) Energy management of WSN with MIMO techniques. Int J Eng Sci Technol 4(6):2534–2541

BER Performance Analysis of Short Reference Differential Chaos Shift Keying Scheme Using Various Maps Over Different Channel Conditions

M. Sangeetha, Toshiba Chamoli and P. Vijayakumar

Abstract In this paper, design and development of chaos-based radio system are realized using short reference differential shift keying. In this communication system, the length of chaos reference sequence is reduced to R so that the message signal frame duration reduces as compared to former DCSK scheme. The proposed system is implemented on AWR-Virtual System Simulator (VSS) tool. VSS software is a comprehensive environment for the end-to-end design and system-level simulation of the communication system. With MATLAB co-simulation, we can readily import MATLAB code into VSS for enhanced system capability. The proposed system will be studied to deduce analytical BER expression under fading channel conditions and validated with simulated and experimental results.

Keywords AWR-VSS · BER analysis · Chaos-based non-coherent modulation Increased data rate · Lower energy utilization · SR-DCSK

1 Introduction

Chaotic codes are non-periodic and random signals generated in a deterministic manner. A slight change in the initial condition for generation of these chaotic codes leads to entirely new chaotic codes which are mutually orthogonal [1]. Chaotic systems have excellent auto-correlation and cross-correlation that gives them an advantage over the conventional communication system. These codes find theirs use in spread spectrum communication systems as a replacement for pseudo-random codes [2].

M. Sangeetha (✉) · T. Chamoli · P. Vijayakumar
Department of ECE, SRM University, Chennai, India
e-mail: sangeetha.m@ktr.univ.ac.in

T. Chamoli
e-mail: toshibachamoli14@gmail.com

P. Vijayakumar
e-mail: vijayakumar.p@ktr.univ.ac.in

A. M. Zungeru et al. (eds.), *Wireless Communication Networks and Internet of Things*, Lecture Notes in Electrical Engineering 493,
https://doi.org/10.1007/978-981-10-8663-2_21

There are different chaos-based communication systems being presented and evaluated; from those systems, differential chaos shift keying (DCSK) is found to be the most robust non-coherent scheme. In [3–5], the performance of various chaos-based communication systems over AWGN and fading channels has been studied. DCSK achieves not only good error performance but also low implementation complexity [6]. In [7], an FM-DCSK system is reviewed and analyzed. In [8], we have a study of the DCSK system with the provision of noise reduction. In this system, noise variance of the demodulated signal is reduced to decrease BER at the receiver.

A reduced reference length DCSK system overcomes low data rate and efficiency of DCSK in [9]. For each information bit, we generate a reference chaos sample consisting of length given as β/P rather than producing β samples and then multiplied P times with the message signal. At the receiver, this reference signal gets demodulated first and then P number of times correlation with its time-delayed version is performed to recover the transmitted signal.

Paper contribution and outline: In this paper, SR-DCSK chaos radio system is studied, analyzed, and implemented over Advanced Wireless Ranging Virtual System Simulator tool. The designing of this modulation scheme is simple without any need for channel estimation at the receiver. This paper can be summarized as follow:

1. Implementing an SR-DCSK system over AWR-VSS tool.
2. Calculating and analyzing BER performance over AWGN channel and fading channel and validating the same with simulation and experimental results.

Organization of this paper as follow: In Sect. 2, we discuss a detailed data transmission methodology of the SR-DCSK system. BER performance analysis of the SR-DCSK scheme is studied in Sect. 3 using different chaos maps. In Sect. 4, we discuss simulation results with various maps and length of the optimal reference signal over AWGN and Rayleigh fading channels, and conclusion is discussed in Sect. 5.

2 SR-DCSK Modulation Scheme

Figure 1a shows that the length of chaotic samples that makes the reference signal is shortened to R samples so that the duration of the frame gets reduced as compared to DCSK. In order to form transmitted information signal, this R length reference signal can be concatenated several times to modulate the data sequence. As a result, this system allows bit duration of $\text{Tb} = R + P * R = R + \beta$. This $R + \beta$ frame length duration of SR-DCSK is less than a conventional DCSK system having 2β frame duration.

Figure 1b represents SR-DCSK transmitter where for each bit b_i, a chaos signal consisting of R number of samples makes the reference signal. This reference signal

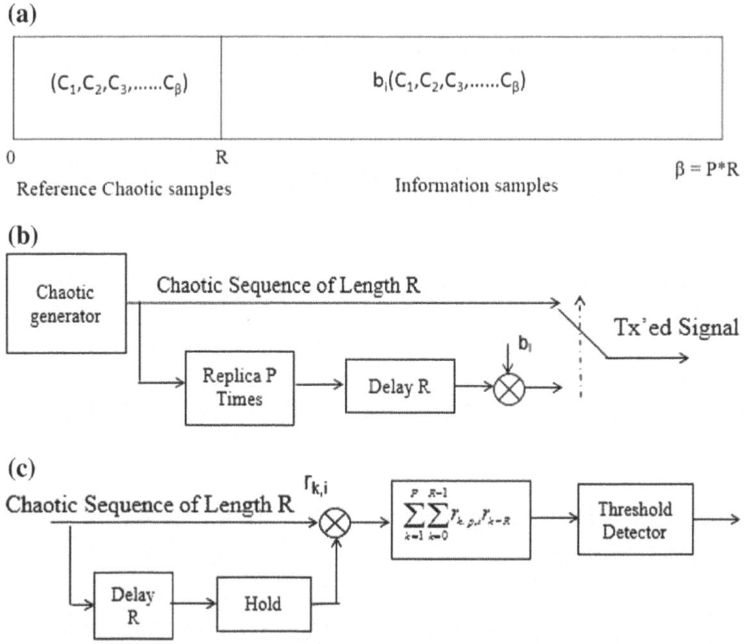

Fig. 1 **a** SR-DCSK frame structure, **b** SR-DCSK transmitter, **c** SR-DCSK receiver

is then used to modulate the data signal over P number of times. The length of the data sequence thus formed is $\beta = P * R$ samples. Now, frame length of an SR-DCSK system that consists of the reference sequence of R samples and a data-modulated signal with β samples where $\beta = P * R$ becomes $(R + \beta)$ or $(R + P * R)$.

In SR-DCSK system, the transmitted signal is given as follow [9]

$$e_{k,i} = \begin{cases} x_{k,i} & \text{for } (0 \leq k < R) \\ b_i x_{k-R,i} & \text{for } (R \leq k \leq (1+P)R \end{cases} \tag{1}$$

In the above equation, $x_{k,i}$ is the chaotic signal, R is reference signal chip length, b_i is information sequence, and P is number of times reference signal gets replicated.

Figure 1c represents SR-DCSK receiver where $r_{k,i}$ represents received signal. In order to extract the reference samples at the receiver, we delay the received signal with a delay of R samples. After the extraction of the reference signal, correlations with remaining data sequence P number of times are carried out over a duration of R samples. The result of the P independent correlations gets added and an estimation of the transmitted bit is carried out by comparing the outcome to a threshold taken at zero. In SR-DCSK system, the received signal given as [9]

$$r_{k,i} = \sum_{l=1}^{L} \alpha_{l,i} e_{k-\tau_{l,i}} + n_k \quad (2)$$

where $a_{l,i}$ and τ_l are the channel gain and delay of lth path, L represents a number of paths, and n_k is AWGN with zero mean and variance No/2.

The output after partial correlation is given as [9]

$$O_{i,p} = T_c \sum_{l=1}^{R-1} \left(\sum_{l=1}^{L} \alpha_{l,i} x_{k-\tau_{l,i}} b_i + n_{p,k+R} \right) \left(\sum_{l=1}^{L} \alpha_{l,i} x_{k-\tau_{l,i}} + n_k \right) \quad (3)$$

where $O_{i,p}$ is the pth partial correlation value, $n_{p,k+R}$ is the kth AWGN noise getting added to the kth sample of the pth data signal sample, and n_k is the AWGN noise.

3 BER Analysis of SR-DCSK System Over Rayleigh Fading Channel

The chaos signal is generated using Logistic Maps for the analysis and is defined as [10]

$$f(x_k) = 1 - 2x_{k-1}^2 \quad (4)$$

For SR-DCSK system, the decision variable can be written as [9]

$$D_i = \sum_{p=1}^{P} O_{i,p} \quad (5)$$

For the ith transmitted bit, the mean of D_i (decision variable) is given by [9]

$$E[D_i] = RPb_l \sum_{l=1}^{L} \alpha_{l,i}^2 E\left[x_{k,i}^2\right] \quad (6)$$

where $RPE\left[x_{k,i}^2\right]$ represents the energy of the received bit from P correlation. Replacing the term $E\left[x_{k,i}^2\right]$ with expression of energy for bit duration of SR-DCSK where $E_b = \frac{RP}{(R+\beta)} b_l \sum_{l=1}^{L} \alpha_{l,i}^2 E_b$, the expression in the above equation becomes

$$E[D_i] = \frac{RP}{R+\beta} b_l \sum_{l=1}^{L} \alpha_{l,i}^2 E_b \quad (7)$$

The variance of D_i is given as [9]

$$V[D_i] = P^2RE\left[x^2_{k-\tau_l,i}\right]\left(\frac{N_0}{2}\sum_{l=1}^{L}\alpha^2_{l,i} + \sum_{l=1}^{L}\sum_{l=1}^{L}\alpha^2_{l,i}\alpha^2_{l,i}E\left[x^2_{k-\tau_l,i}\right]\right) + PR\frac{N_0^2}{4} + PR\frac{N_0}{2}\sum_{l=1}^{L}\alpha^2_{l,i}E\left[x^2_{k-\tau_l,i}\right] \quad (8)$$

From Eqs. 7 and 8, the BER for SR-DCSK system may be defined as [9]

$$\text{BER} = \frac{1}{2}erfc\left(\left[\frac{(R+\beta)N_0}{R\sum_{l=1}^{L}\alpha^2_{l,i}E_b}\left(\frac{P+1}{P}\right) + \frac{(R+\beta)^2N_0^2}{2RP\left(\sum_{l=1}^{L}\alpha^2_{l,i}\right)E_b^2}\right]^{-\frac{1}{2}}\right) \quad (9)$$

Equation 9 shows theoretical BER over channels undergoing multipath fading. Since channel keeps on changing at every instant of time, we apply an average lower bound to equation [9]

$$\text{BER} = \frac{1}{2}erfc\left(\left[\frac{(R+\beta)}{\gamma}\left(\frac{P+1}{P}\right) + \frac{(R+\beta)^2}{2RP\gamma^2}\right]^{-\frac{1}{2}}\right)f(\gamma)\text{d}\gamma \quad (10)$$

where $\gamma = \sum_{l=1}^{L}\alpha_l^2\frac{E_b}{N_o}$

For L independent and identically distributed Rayleigh fading channel, the PDF of γ is given as [11]

$$f(\gamma) = \frac{\gamma^{L-1}}{(L-1)!\gamma^L}\exp\left(-\frac{\gamma}{\gamma_c}\right) \equiv f(\gamma, \gamma_{c,}, L) \quad (11)$$

where $\gamma_c = \frac{E_b}{N_0}E\left(\alpha_j^2\right) = \frac{E_b}{N_0}E(\alpha_l^2),\ j \neq l$

Analytical BER expression of SR-DCSK system in an AWGN channel is given as [9]

$$\text{BER} = \frac{1}{2}erfc\left(\left[\frac{(R+\beta)N_0}{RE_0}\left(\frac{P+1}{P}\right) + \frac{(R+\beta)^2N_0^2}{2RPE_0^2}\right]^{-\frac{1}{2}}\right) \quad (12)$$

In the above BER expression,

$\frac{E_b}{N_0}$ SNR
R Reference chaotic sequence length
P Repetition factor
β Spreading length ($P \times R$)

4 Simulation Result

AWR-VSS tool is used to simulate and analyze SR-DCSK system performance. VSS software helps the user to design and analyze end-to-end communication system. Based on our analysis, we can efficiently display our result as BER curves, constellation, and power spectrum, etc. This software enables users to readily import MATLAB code into VSS for enhanced system simulation capability.

In order to generate the waveform for chaos signal, this paper uses Logistic Map with initial condition taken as 0.134. The length of the reference signal is 15 which is repeated 20 times in order to form the transmitted frame of length 300 chips (Figs. 2, 3 and 4).

As observed from Figs. 5, 6, and 7 for SR-DCSK modulation scheme, BER versus SNR graph is a waterfall curve, and as SNR increases BER decreases. From above graphs, we note that Logistic Map gives better BER performance as compared to Bernoulli and Tent Map.

We compute the analytical BER for a spreading length of $\beta = 300$, with $P = 20$ and $R = 15$. The BER computed for SNR (dB) ranging from 0 to 20. Now, a comparative study is done between the analytical and simulated BER to validate the simulated result obtain for different values of SNR for different chaotic maps. Graph obtained for the same is shown below.

In the above graph, BER plot for three different cases—Analytical BER, simulated BER for Logistic Map, and simulated BER for Bernoulli Shift Map—is obtained. Above BER values are calculated considering spreading length $\beta = 300$, $P = 20$ and $R = 15$. The value of SNR ranges from 0 to 20 dB. The above graph shows that simulation value of BER for the Logistic Map is close to analytical value, whereas for Bernoulli Shift Map, there is much deviation from the analytical BER value. Thus, we can say that Logistic Maps provide better performance as compared to Bernoulli Shift Map.

Figure 7 represents the performance of SR-DCSK scheme in a Rayleigh fading channel with $L = 1$. We observe the behavior of the system by varying the reference

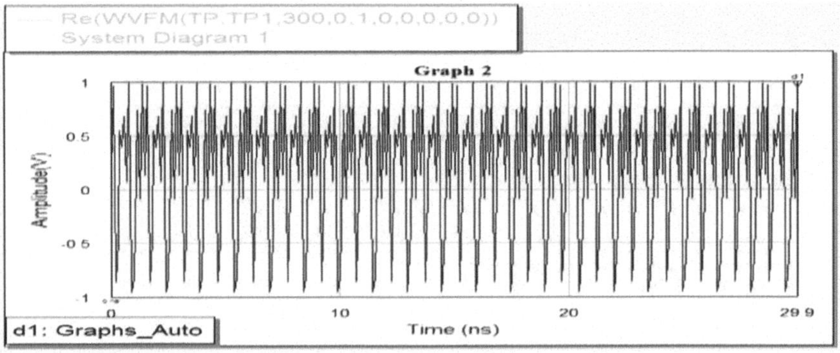

Fig. 2 Waveform of chaotic signal generated using Logistic Map in AWR-VSS tool

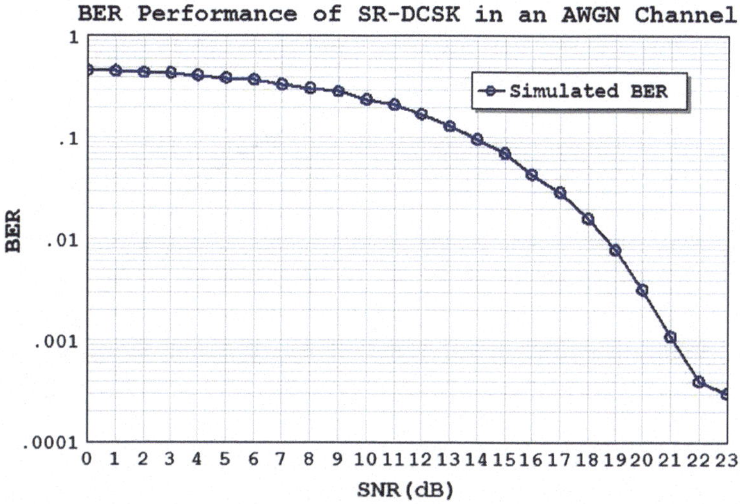

Fig. 3 Simulated BER result of SR-DCSK system through AWR-VSS tool using Logistic Map

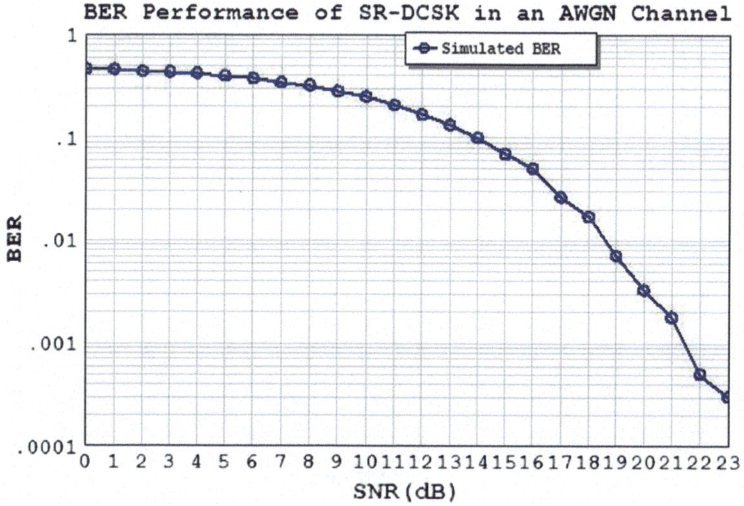

Fig. 4 Simulated BER result of SR-DCSK system through AWR-VSS tool using Bernoulli shift map

signal length for various values of *R*. As shown in the graph, for higher value of reference length, BER decreases. Figure 8 verifies the performance of SR-DCSK for Rayleigh fading channel using AWR-tool.

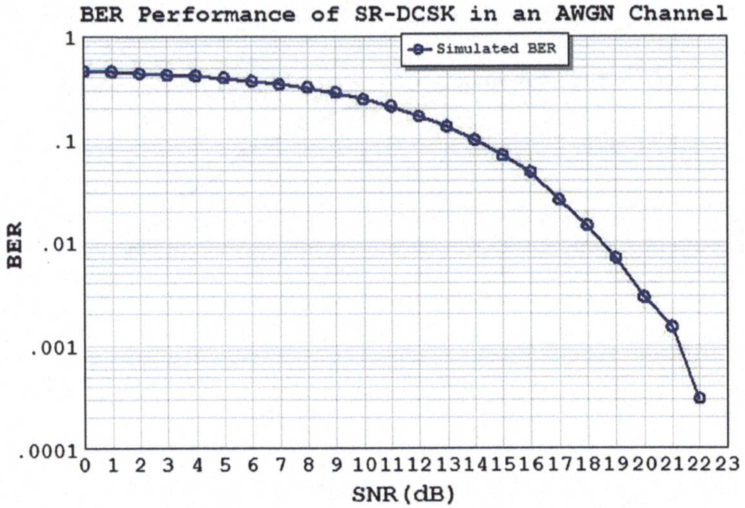

Fig. 5 Simulated BER result of SR-DCSK system through AWR-VSS tool using tent map

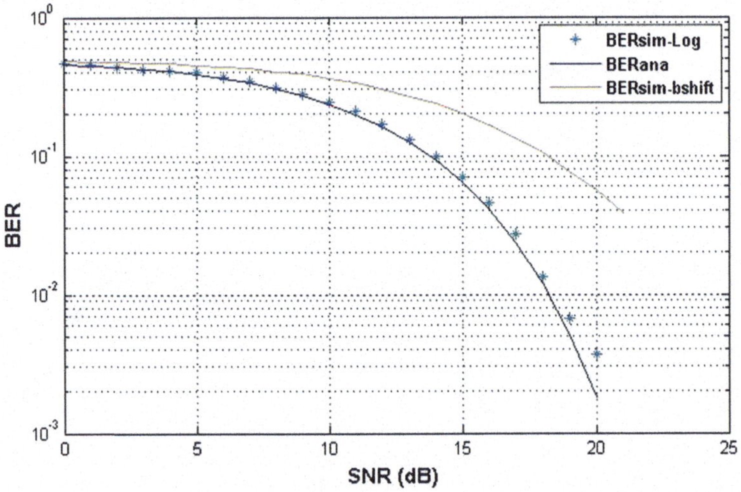

Fig. 6 Analytical versus Simulated BER graph for SR-DCSK scheme over AWGN channel using MATLAB

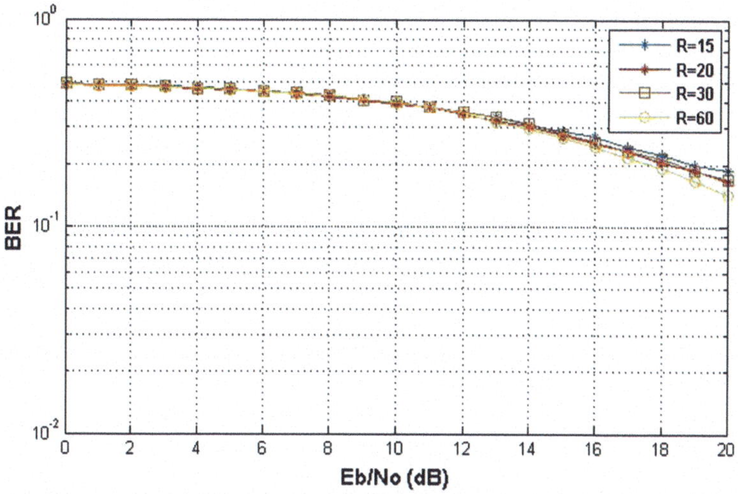

Fig. 7 BER graph for SR-DCSK scheme over Rayleigh fading channel for R = 15, 20, 30, 60 using MATLAB

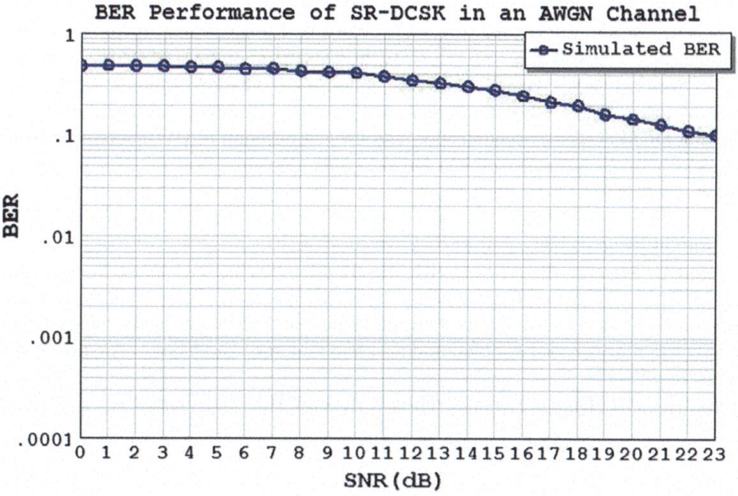

Fig. 8 BER graph for SR-DCSK scheme over Rayleigh fading channel for R = 60 using AWR-VSS tool

5 Conclusion

This paper studies the performance of the SR-DCSK system. SR-DCSK is more energy efficient when compared to DCSK because of short reference signal duration. As a result, it provides an additional advantage of higher data rate over DCSK.

The implementation of this system is easy because of simplicity in designing. The performance of the SR-DCSK system is analytically studied and simulated using MATLAB and AWR-VSS tool to verify the BER expression under AWGN and Rayleigh fading channel. Moreover, the effect of a change in length of the reference signal is studied and analyzed, and we observed that the BER of the system decreases as we increase reference signal length.

Acknowledgements The simulation tool of AWR-VSS is sponsored by AWR-A National Instrument Company, under the NI–AWR sponsored research project, May 2016.

References

1. Kaddoum G, Member IEEE (2013) Wireless chaos based communication system: a comprehensive study. IEEE
2. Lau FCM, Tse CK (2003) Chaos-based digital communication systems. Springer, Berlin
3. Kennedy MP, Kolumbán G, Kis G, Jákó Z (2000) Performance evaluation of FM-DCSK modulation in multipath environments. IEEE Trans Circuits Syst 47:1702–1711
4. Kaddoum G, Chargé P, Roviras D (2009) A generalized methodology for bit-error-rate prediction in correlation-based communication schemes using chaos. Commun Lett 13 (8):567–569
5. Sushchik M, Tsimring LS, Volkovskii AR (2000) Performance analysis of correlation-based communication schemes utilizing chaos. IEEE Trans Circuits Syst 47:1684–1691
6. Fang Y, Member, IEEE, Han G, Senior Member, IEEE, Chen P, Member, IEEE, Lau FCM, Senior Member, IEEE, Chen G, Fellow, IEEE, Wang L, Senior Member, IEEE (2015) A survey on DCSK based communication and their application in UWB scenarios. IEEE
7. Chen S (2016) Performance of FM-DCSK communication systems with timing synchronization error. IEEE
8. Kaddoum G, Member, IEEE, Soujeri E, Senior Member, IEEE (2016) NR-DCSK: a noise reduction differential chaos shift keying system. IEEE
9. Kaddoum G, Member, IEEE, Soujeri E, Senior Member, IEEE, Nijsure Y, Member, IEEE (2016) Design of a short reference non-coherent chaos-based communication systems. IEEE
10. Kaddoum G, Chargé P, Roviras D, Fournier-Prunaret D (2009) A methodology for bit error rate prediction in chaos-based communication systems. Birkhäuser Circuits Syst Signal Process 28:925–944
11. Proakis JG (2001) Digital communications. McGraw-Hill, New York

NR-DCSK-Based MIMO Chaotic Communication System

Sangeetha Manoharan, Niharika Saraff, Akanksha Kedia and Kasturi Laxmi Saroja

Abstract In this paper, noise reduction differential chaos shift keying (NR-DCSK)-based multiple-input–multiple-output (MIMO) chaotic communication system is proposed. We consider a transceiver scheme that uses a different chaotic sequence at each transmit antenna and transmits omnidirectionally. In the transmitter side, we use β/P chaotic samples as reference sequence which is then repeated P times to generate the required reference sequence of length β. At the receiver side, a moving average filter of size P is used to reduce the noise variance and improve the efficiency of the system. The resulting filtered signal is then correlated and combined with equal gain combining to yield the transmitted bit. Bit error rates for additive white Gaussian noise (AWGN) channel are analytically derived. These results are compared with the simulated results to validate the expressions derived. Results show that NR-DCSK-based MIMO system is better than the conventional M-DCSK-based chaotic communication system.

1 Introduction

Chaotic signals are non-periodic, random signals which are derived from nonlinear dynamical systems [1–4]. Chaotic modulation is employed in a number of applications such as spread spectrum communication, radar and optical communications. The advantages of chaotic signals are that they are sensitive to initial conditions and have low complexity in hardware implementation and consume less power compared to traditional communication methods. Among the various chaotic modulation schemes, differential chaos shift keying (DCSK) is the most suitable for wireless communication as it is robust and non-coherent in nature [4].

In DCSK, the entire frame is divided into two equal lengths consisting of the reference chaotic signals slot and the data-bearing slot. If +1 is transmitted, then the data sample follows the reference signal, and if −1 is transmitted, then the data

S. Manoharan (✉) · N. Saraff · A. Kedia · K. L. Saroja
ECE Department, SRM University, Kattankulathur, India
e-mail: sangeetha.m@ktr.srmuniv.ac.in

A. M. Zungeru et al. (eds.), *Wireless Communication Networks and Internet of Things*, Lecture Notes in Electrical Engineering 493,
https://doi.org/10.1007/978-981-10-8663-2_22

sample is an inverted copy of the reference signal. The spreading factor in a DCSK system is denoted by 2β which represents the number of chaotic samples. The noise addition to both the reference sequence and the data-bearing sequence is a major disadvantage of the DCSK system as it diminishes the performance of the system.

To overcome this drawback, we go for an enhanced version of DCSK known as noise reduction differential chaos shift keying (NR-DCSK) [1] system. In NR-DCSK system, we generate β/P different chaotic samples rather than β samples to be utilized as reference signal. These β/P different samples are then replicated by a factor P in the signal. At the receiver side, a moving average filter is averaged over a window size P to reduce the noise variance [1]. This filtered signal is then correlated with equal gain combining to yield the transmitted signal.

Multiple-input–multiple-output (MIMO) is an antenna configuration which enhances the capacity of a wireless system. The transmitted and the receiving antennas are configured in a way to minimize the errors. In this brief, we propose a NR-DCSK-based MIMO chaotic communication system where the NR-DCSK modulation scheme is employed in a wireless MIMO system and its performance is analyzed for various scenarios.

The remaining part of this brief is organized as follows: Section 2 describes the architecture of the system. The BER performance of the proposed NR-DCSK-based MIMO system is presented in Sect. 3. Section 4 represents the simulation results and the inference derived from it. Section 5 gives the conclusion of the brief.

2 System Model

2.1 NR-DCSK Modulation

In NR-DCSK modulation, β/P chaotic samples are generated, where P is the reduction factor and β is the spreading factor. The chaotic samples that are generated are replicated by a factor P to get the reference signal which has a length β. Similar to M-DCSK, in NR-DCSK modulation scheme the transmitted bit is represented by two chaotic signals: reference signal and data-bearing signal. The transmitted signal is represented as $e_{i,k}$ [1],

$$e_{i,k} = \begin{cases} x_{i,\lceil \frac{k}{P} \rceil}, & 0 < k \leq \beta \\ b_i x_{i,\lceil \frac{k}{P} \rceil - \beta}, & \beta < k \leq 2\beta \end{cases} \tag{1}$$

where $x_{i,\lceil \frac{k}{P} \rceil}$ is the reference sequence and $x_{i,\lceil \frac{k}{P} \rceil - \beta}$ is its delayed version, b_i is the information bit, and $\lceil \Lambda \rceil$ represents the ceiling operator. Equation (1) represents the transmitted signal for a single-input–single-output (SISO) system. At the receiver side, a moving average filter is used to average the P samples which are identical. This filtered signal is then correlated with its delayed replica so as to yield the bit transmitted [1].

2.2 *NR-DCSK-Based MIMO System*

The paper focuses on employing NR-DCSK modulation scheme in a wireless MIMO system. The frame structure of NR-DCSK system is illustrated in Fig. 1.

In the proposed MIMO system, there are N_t transmitting antennas and N_r receiving antennas over an AWGN channel. The transmitted signal of NR-DCSK MIMO system for *i*th bit is given in Eq. (2) (Fig. 2)

$$s_{i,k} = \begin{cases} \sqrt{\frac{\rho}{N_t}} x_{i,\lceil \frac{k}{P} \rceil}, & 0 < k \leq \beta \\ \sqrt{\frac{\rho}{N_t}} b_i x_{i,\lceil \frac{k}{P} \rceil - \beta}, & \beta < k \leq 2\beta \end{cases} \tag{2}$$

where ρ is the total transmitted signal-to-noise ratio (SNR) [2].

The received signal at *j*th receive antenna is represented by Eq. (3)

$$r_{j,k} = \sum_{i=1}^{N_t} \left(\left| h^0_{j,i} \right| s_{i,k} + \left| h^1_{j,i} \right| s_{i,\lceil \frac{k}{P} \rceil - \tau} \right) + \text{noise} \tag{3}$$

where $h^0_{j,i}$ and $h^1_{j,i}$ are the channel coefficients and τ is the time delay. The received signal is then averaged using a moving average filter with a window size *P*. This averaged signal is then correlated with its time-delayed version which is delayed by a factor β/P which is represented as follows:

$$U_{j,k} = \sum_{k=1}^{\frac{\beta}{P}} r_{j,k} r^*_{j,\lceil \frac{k}{P} \rceil - \tau} \tag{4}$$

The outputs of the correlators are then combined with equal gain combining (EGC) technique [2] which is given as

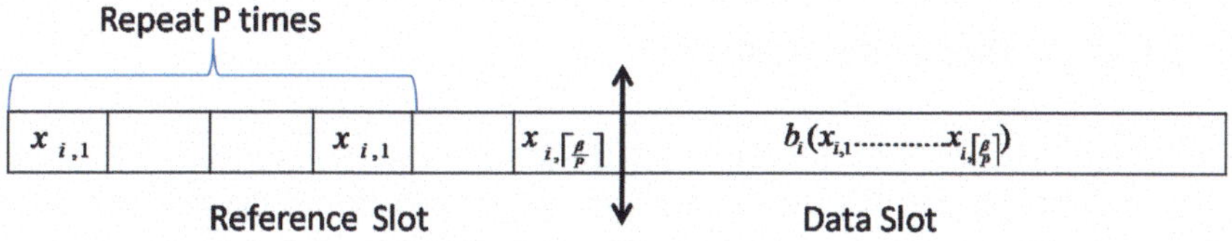

Fig. 1 MIMO NR-DCSK frame structure

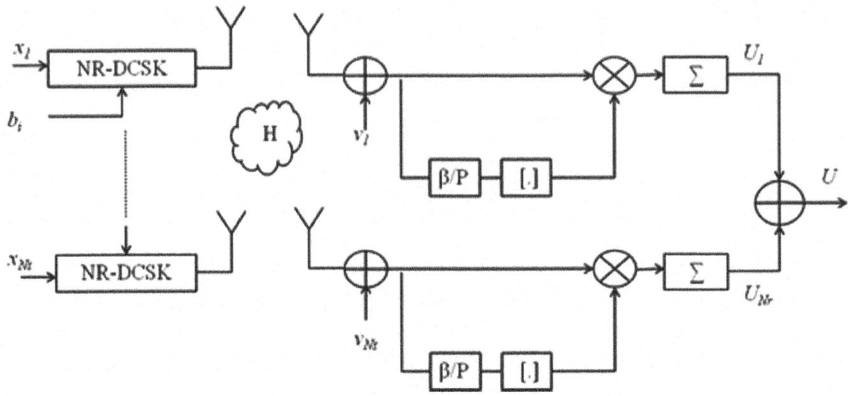

Fig. 2 Block diagram of NR-DCSK based MIMO system model

$$
\begin{aligned}
U_{j,k} = \sum_{k=1}^{\frac{\beta}{P}} \sum_{i=1}^{N_t} \sum_{j=1}^{N_r} \Bigg[& \left|h_{j,i}^{0}\right|^2 \frac{\rho}{N_t} b_i x_{i,\lceil \frac{k}{P} \rceil}^2 + \left|h_{j,i}^{1}\right|^2 \frac{\rho}{N_t} b_i x_{i,\lceil \frac{k}{P} \rceil - \tau}^2 \\
& + \frac{2}{N_t} b_i \left|h_{j,i}^{0}\right| \left|h_{j,i}^{1}\right| \rho x_{i,\lceil \frac{k}{P} \rceil} x_{i,\lceil \frac{k}{P} \rceil - \tau} \\
& + \left(\left|h_{j,i}^{0}\right| x_{i,\lceil \frac{k}{P} \rceil} + \left|h_{j,i}^{1}\right| x_{i,\lceil \frac{k}{P} \rceil - \tau} \right) \sqrt{\frac{\rho}{N_t}} \frac{1}{P} \sum_{p=1}^{P} n_{i,p+k+\frac{\beta}{P}} \\
& + \left(\left|h_{j,i}^{0}\right| b_i x_{i,\lceil \frac{k}{P} \rceil} + \left|h_{j,i}^{1}\right| b_i x_{i,\lceil \frac{k}{P} \rceil - \tau} \right) \sqrt{\frac{\rho}{N_t}} \frac{1}{P} \sum_{p=1}^{P} n_{i,p+k} \\
& + \frac{1}{P} \sum_{p=1}^{P} n_{i,p+k} \frac{1}{P} \sum_{p=1}^{P} n_{i,p+k+\frac{\beta}{P}} \Bigg]
\end{aligned} \tag{5}
$$

where first term represents the useful message component, second term represents the interference component, and $\frac{1}{P}\sum_{p=1}^{P} n_{i,p+k}$ represents the noise component after it passes through the moving average filter [1]. $n_{i,k}$ is AWGN and has a mean zero and variance No/2.

3 BER Performance Analysis

In this part, the performance of the NR-DCSK-based MIMO system is evaluated under AWGN channel. We then find the mean and variance of the decision variable, $U_{j,k}$, in Eq. (5). The mean of the ith transmitted bit is given as

$$E\left[U_{j,k}\right] = \frac{\beta\rho}{PN_t}\left(\left|h^0_{j,i}\right|^2 + \left|h^1_{j,i}\right|^2\right)E\left[x^2_{i,k}\right] \tag{6}$$

where $E[.]$ denotes the expected value operator. The variance of the decision variable is given as

$$\begin{aligned}\mathrm{Var}\left[U_{j,k}\right] &= \frac{\beta\rho}{PN_t}\left(\left|h^0_{j,i}\right|^2 + \left|h^0_{j,i}\right|^2\right)\mathrm{Var}\left[x^2_{i,k}\right] \\ &+ \sqrt{\frac{\rho}{N_t}\frac{N_o}{P}\frac{\beta}{P}}E\left[x^2_{i,k}\right] + \frac{\beta N_o^2}{4P^3}\end{aligned} \tag{7}$$

From Eqs. (6) and (7), we can observe that the mean of decision variable in Eq. (5) is equal to the mean of only the useful message component. All the other components have zero mean for large β/P values as

$$\sum_{k=1}^{\frac{\beta}{P}} \left(x_{k-\tau}x_k\right) \approx 0$$

The output of the correlator follows Gaussian distribution, and therefore, the bit error rate probability is represented as [4]

$$\mathrm{BER} = \frac{1}{2}\Pr(U_i < 0 | b_i = +1) + \frac{1}{2}\Pr(U_i > 0 | b_i = -1) \tag{8}$$

which is represented as complementary error function in the following way

$$\mathrm{BER} = \frac{1}{2}\mathrm{erfc}\left(\left[\frac{2\mathrm{Var}[U_i]}{E[U_i]^2}\right]^{\frac{-1}{2}}\right) \tag{9}$$

erfc(z) is defined as $\mathrm{erfc}(z) = (2/\sqrt{\pi})\int_z^{\infty} e^{-u^2}du$

By combining Eqs. (6), (7) and (9) the following BER expression is obtained

$$\mathrm{BER} = \frac{1}{2}\mathrm{erfc}\left[\frac{4}{\left(\frac{\rho}{N_t}\right)^{\frac{3}{2}}\gamma} + \frac{2\beta}{P\left(\frac{\rho}{N_t}\right)\gamma_2} + \frac{2\Psi}{\left(\frac{\beta}{P}\right)\left(\frac{\rho}{N_t}\right)\left(\left|h^0_{j,i}\right|^2 + \left|h^1_{j,i}\right|^2\right)}\right] \tag{10}$$

where E_b is the transmitted bit energy, and γ and ψ are represented as

$$\gamma = \left(\left| h_{j,i}^{0} \right|^2 + \left| h_{j,i}^{1} \right|^2 \right) \frac{E_b}{N_o} N_r$$

$$\psi = \frac{\mathrm{Var}\left[\left(x_{i,k} \right)^2 \right]}{E^2 \left[\left(x_{i,k} \right)^2 \right]}$$

In an AWGN channel, the channel coefficients are chosen as $\left| h_{j,i}^{0} \right| = 1$ and $\left| h_{j,i}^{1} \right| = 0$.

4 Simulation Results

To analyze and validate the BER performance of NR-DCSK scheme in wireless MIMO systems, it is compared with the BER performance of conventional M-DCSK scheme in wireless MIMO system. Both simulation and analytical results are illustrated in this section. Figure 3 illustrates the sequence structure of the Logistic Map.

When computing the BER for NR-DCSK-based MIMO system by using different maps for chaos signal generation, the BER performance is almost the same. Thus, NR-DCSK-based MIMO system is so robust that irrespective of the map used for chaos generation, the performance remains unaffected.

The BER expressions of NR-DCSK scheme for various antenna configurations which are single-input–single-output (SISO) system, single-input–multiple-output

Fig. 3 Sequence structure for Logistic Map

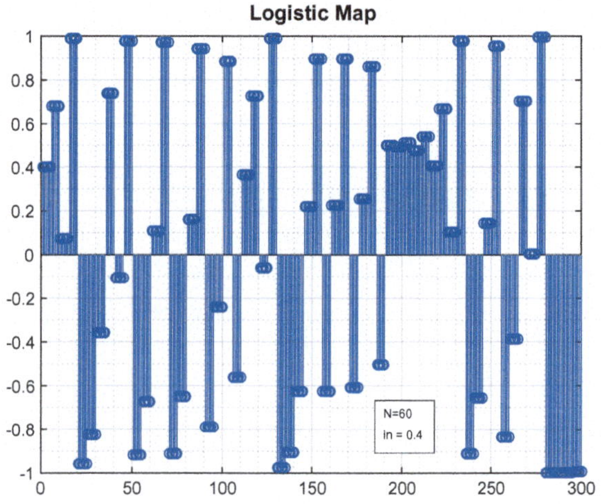

(SIMO) system, multiple-input–single-output (MISO) system, and multiple-input–multiple-output (MIMO) system were also computed. Figure 4 shows the BER performance of the NR-DCSK scheme for the above-mentioned configurations.

Figure 4 shows that SIMO system yields the best performance, followed by MIMO system. Thus, the performance of the system improves as the number of antennas on the receiver side increases.

Figure 5 illustrates the simulation and analytical BER performance of M-DCSK-based MIMO system and NR-DCSK-based MIMO system. It perfectly validates the accuracy of the analytical BER expression derived and mentioned as Eq. 10 in the paper. Figure 5 also shows that NR-DCSK scheme is better than M-DCSK scheme for MIMO systems. For example, at 5 dB SNR, the BER for M-DCSK-based MIMO system is 0.2790, while the BER for NR-DCSK-based MIMO system is 0.039. This clearly shows NR-DCSK-based MIMO system gives enhanced performance.

On examining Fig. 6, we can say that higher the value of reduction factor, i.e., P, better is the performance of the system. When $P = 1$, then the system reduces to the conventional M-DCSK system, and hence, the performance of the NR-DCSK scheme is better than M-DCSK scheme [1]. Thus, we can say that increase in value of P decreases the noise variance of the NR-DCSK system, thereby enhancing the performance.

Thus, the overall performance of NR-DCSK-based MIMO systems increases with increase in the value of reduction factor and also with the increase in number of receiving antennas. Thus, NR-DCSK-based MIMO system yields better signal-to-noise ratio (SNR) than the conventional M-DCSK-based MIMO system. The system parameters are given in Table 1.

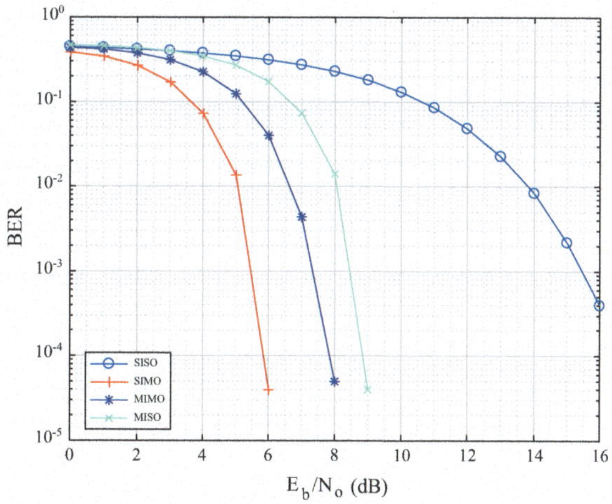

Fig. 4 Analytical BER performance analysis for various configurations of antenna using NR-DCSK scheme

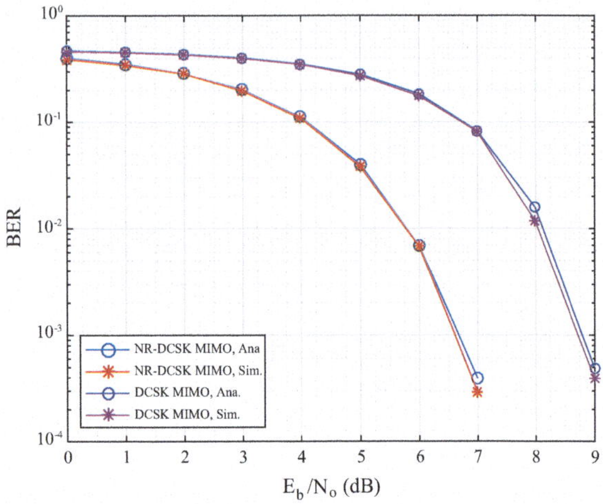

Fig. 5 Simulation and analytical comparison of NR-DCSK-based MIMO scheme with conventional M-DCSK-based MIMO system

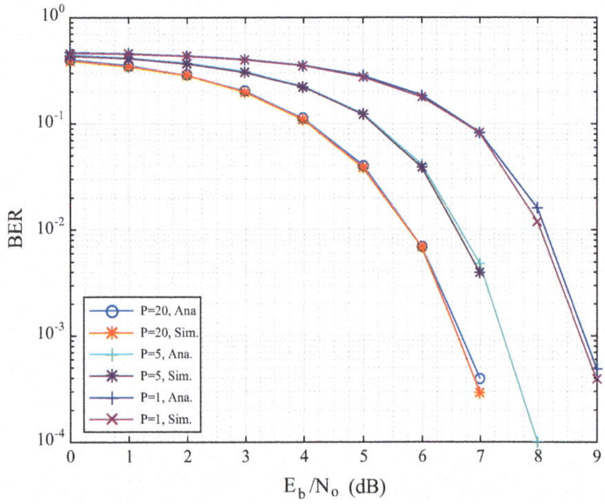

Fig. 6 Simulation and analytical results for NR-DCSK-based MIMO system for P = 1, 5 and 20

Table 1 System parameters

Symbol	Representation	Quantity
M	Number of information bits	1000
B	Spreading factor	300
P	Replication factor	20
β/P	Reduction factor	15
T_c	Chip duration	1
βT_c	Bit duration	300
SNR	Signal-to-noise ratio	0–20 dB

5 Conclusion

A NR-DCSK-based MIMO chaotic communication system is proposed in this brief. The performance of the proposed NR-DCSK-based MIMO system is analytically studied, and the BER expression is then derived under AWGN channel. The validity of the analytical BER expression is confirmed by carrying out computer simulation. The results show that NR-DCSK-based MIMO systems yield higher SNR and lower BER than the conventional M-DCSK-based MIMO system and the system performance enhances with increase in the value of reduction factor.

References

1. Kaddoum G, Soujeri E (2016) NR-DCSK: a noise reduction differential chaos shift keying system. IEEE Trans Circ Syst II Express Briefs 63(7):648–652
2. Wang S, Wang X (2010) M-DCSK-based chaotic communications in MIMO multipath channels with no channel state information. IEEE Trans Circuits Syst II Express Briefs 57(12):1001–1005
3. Thapaliya K, Yang Q, Kwak KS (2007) Chaotic communications in MIMO systems. In: Lee Y-H et al (eds.) ICESS 2007, LNCS 4523. Springer, Berlin, Heidelberg, pp 708–717
4. Lau FCM, Tse CK (2003) Chaos based digital communication systems. Springer, Berlin, Germany

Wearable Sensor-Based Human Fall Detection Wireless System

Vaishna S. Kumar, Kavan Gangadhar Acharya, B. Sandeep, T. Jayavignesh and Ashvini Chaturvedi

Abstract *Background/Objectives*: Human fall detection is a critical challenge in the healthcare domain since the late medical salvage will even lead to death situations, therefore it requires timely rescue. This research work proposes a system which uses a wearable device that senses human fall and wirelessly raises alerts. *Methods/statistical analysis*: The detection system consists of the sensor system which contains both accelerometer and gyroscope sensors. The proper orientation of the subject is provided by the Madgwick filter. Six volunteers were engaged to perform the falling and non-falling events. The system is validated and checked by four algorithms: threshold based, support vector machine (SVM), *K*-nearest neighbor, and dynamic time wrapping, and thus, the accuracy was calculated. *Findings*: From the results obtained, the SVM has given an accuracy of 93%. *Conclusions*: When a fall is being detected, an additional feature to check whether the person is in critical state and is lying down for more than a particular time is incorporated and a critical alert is sent to the caretaker's mobile.

Keywords Accelerometer · Dynamic time wrapping · Gyroscope *K*-nearest neighbor · Madgwick filter · Support vector machine

---

V. S. Kumar · T. Jayavignesh (✉)
School of Electronics Engineering, VIT University, Chennai, Tamil Nadu, India
e-mail: jayavignesh.t@vit.ac.in

V. S. Kumar
e-mail: vaishna.skumar@gmail.com

K. G. Acharya · A. Chaturvedi
Department of Electronics and Communication Engineering, NITK, Suratkal, Karnataka, India
e-mail: kavanaks@gmail.com

A. Chaturvedi
e-mail: chadurvediashvini@gmail.com

B. Sandeep
ETP3, Robert Bosch Engineering and Business Solutions Private Limited, Bangalore, India
e-mail: Sandeep.B@in.bosch.com

A. M. Zungeru et al. (eds.), *Wireless Communication Networks and Internet of Things*, Lecture Notes in Electrical Engineering 493,
https://doi.org/10.1007/978-981-10-8663-2_23

1 Introduction

As the world's aging population is increasing day by day in this twenty-first century, the research in multidisciplinary areas like health care and monitoring systems has increased tremendously. Most of the injuries in hospital for elderly people above 65 years [1] are due to fall events. Fall events usually lead to some fractures, and if not attended on time, may even lead to death [2]. Unfortunately, the rate of injuries due to fall increases because of the late medical salvage to the affected person as the fall is unnoticed on time to others in home.

Leaving elderly people alone in home and other care centers have increased so much nowadays, and it is leading to many nonfatal injuries. According to the previous studies [3], the impact of the fall depends on the age. The elderly people will not be in a position to seek help from anyone as they may fall into unconscious state or there would not be any people nearby which thereby increases the injury cases. However, due to the availability of the low-cost precise sensors and wireless ad hoc networks, this problem can be approached in a different perspective.

Many researches to detect fall among elderly are being continuously done over past several years. Different approaches have been inferred to detect the fall, which mainly includes the wearable sensors which include the use of only accelerometer or combination of accelerometer and gyrometer sensors [4] and (or) magnetometer added to it. The magnetometer is mostly avoided as the device is generally being used inside home and the magnetic fields from other home appliances may interfere resulting in false results.

One of the earlier systems which was proposed was to detect the falls automatically with the help of passive infrared sensors (PIRs) and pressure mats (PMs) [5] with the help of a wireless sensor network and by reading the room coordinates and the movement of the subject, and if the person falls and is not getting up for certain time, alert will be sent. But this is not as effective, as it can be implemented only for a particular room.

The video [6] detection systems are most commonly used at present systems. But this kind of solutions is limited to the area under observation, and the accuracy will be less because most of the falls happen during night time [7] or in bathroom where ubiquitous monitoring using video surveillance is not effective.

Typical changes in the output of a three-axis accelerometer [8] can be used to detect the fall. Simple accelerometer will give sudden spikes in the graph when a fall occurs. Considering the quaternion and thereby calculating the Euler angles to find the orientation of the subject's body with respect to earth, better results can be obtained. Better Euler angles can be obtained by the combination of gyrometer and accelerometer [9].

The fusion of the sensor data to find the Euler angles (roll, pitch, and yaw) is found to be effectively done using Madgwick filter [10] compared with other filters like complimentary filter, Kalman filter [11].

Threshold-based techniques [12] have been mostly used for finding the falling events which will mainly to check whether the RMS value of acceleration and angle has crossed beyond the already set threshold value. For better result, if the values are not changing for certain period of time about 400–500 ms [13], then the person is actually in need of help as he is not able to get up and it is a real fall.

Some products by companies like Masimo are available in market where the sensors are implemented in wristwatch [14], and when a fall is detected, the alarm can be manually operated or the GPS location will be sent to the caretaker using an RF [15] or ZigBee module [16]. The manual operation [17] of alarm is not an effective way in the case when the elderly fell unconscious after the fall.

Many statistical analyses were carried out using the embedded sensors in the mobile phones [18]. In this research, mainly four types of algorithms are being discussed: threshold based, support vector machine (SVM) [19–21], *K*-nearest neighbor (*K*-NN) [22], and dynamic time wrapping (DTW). The trials [23] were carried out in all the four techniques, and the accuracy [24] for all was computed [25–27].

The BMI160 Bosch Sensor which contains both the accelerometer and the gyroscope was used for getting the sensor values, and sensor fusion was done by the Madgwick filter, and further processing is done inside a development kit with EFM32 microcontroller which is a Bosch product, and the alert is sent through the Wi-fi module, and the alert message is being received on a mobile phone and received through an android application called "UDP Sender/Receiver."

Various cases were taken into account so that wrong alert can be avoided while training the SVM and other techniques. The various cases include various daily activities like walking, jogging, running, sitting, bending, picking something from floor, leaning toward wall, squatting, climbing stairs up and down, trembling, slipping, sleeping on floor, and some random activities. The falling events include falling front, back, side, falling from cot, falling from chair, falling from step.

2 Proposed Work

2.1 System Architecture

Figure 1 shows the simplified system architecture of the device used for fall detection.

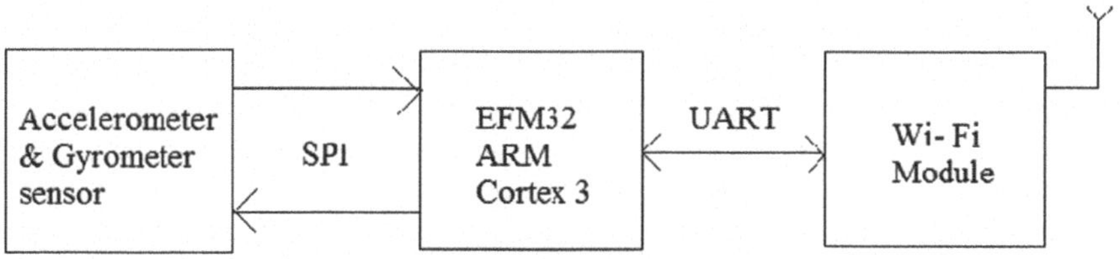

Fig. 1 System architecture

2.1.1 Sensor System

The sensor used is BMI160 which has both 16-bit triaxial accelerometer and gyroscope sensor embedded in it. It provides precise acceleration and angular rate measurement. Gyroscope is also used since it will give the orientation with respect to the earth's gravity. Accelerometer will measure only the non-gravitational acceleration. If only accelerometer is used, the chances of getting false alert are more.

2.1.2 Processor and Wi-fi Module

A Bosch development kit which has BMI160 sensor, a controller (EFM32), and inbuilt Wi-fi module in it, is used for data acquisition, implementing the algorithm, and testing the results.

2.1.3 Receiver

The result sent through Wi-fi module is being received in an open-source application in android "UDP Sender/Receiver." The datasets can be saved in the memory card which is inbuilt in the development kit to do further processing if needed. The kit and the receiver mobile should be connected to the same Wi-fi network.

3 Block Diagram

Figure 2 shows the block diagram of the whole fall detection system.

3.1 Orientation Measurement of the Human Body (Three Dimensional)

Other than the acceleration values, the accurate orientation measurement of the human body plays an important role in finding the fall. To find the posture of human body, the Euler angles have to be computed, which includes the three elemental rotations in all the three axis *X*, *Y*, and *Z* which is represented as roll, pitch, and yaw angles. The basic reference system [10] used is shown in Fig. 3. Thus, the computation of proper roll, pitch, and yaw will give the correct orientation of the human posture which is being set as a threshold to find whether the person has fallen down or not by calculating the orientation with respect to the earth.

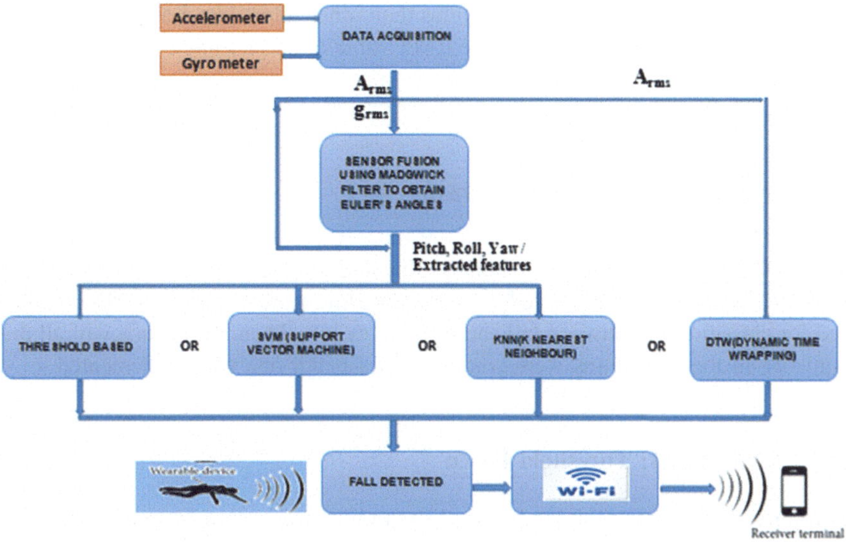

Fig. 2 Block diagram for fall detection

Fig. 3 Basic reference system for Euler angles

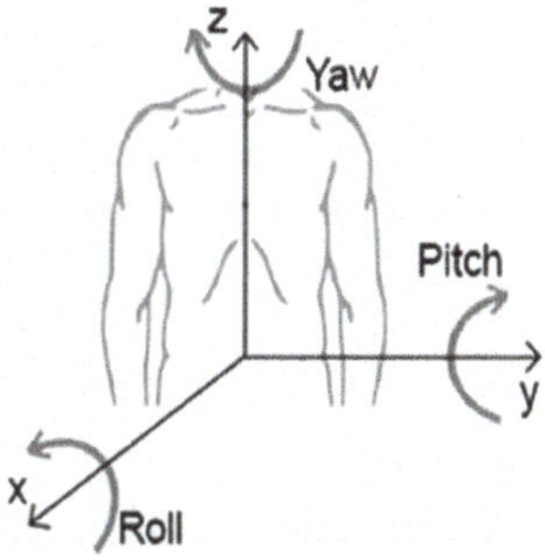

The device is placed on the waist of the subject, and the orientation of the body is calculated from the received data. If the tilt is measured only using accelerometer data, it will be affected by external accelerations and vibrations thus giving false results. Thus to find the orientation, the gyroscope is also included.

3.2 *Sensor Fusion*

The Euler angles can be obtained individually from accelerometer and gyrometer. For finding the Euler angles, better quaternion has to be obtained as it gives the rate of change of earth frame relative to the sensor frame. For getting better angles, sensor fusion has to be done. Sensor fusion is done using Madgwick filter which was found to be giving better Euler angles than other filters like Kalman filter. The input to the Madgwick filter [28] is the RMS value of accelerometer and gyrometer data.

3.3 Algorithm for Fall Detection

Mainly four types of algorithms, namely: threshold based, support vector machine (SVM), *K*-nearest neighbors (*K*-NN), and dynamic time wrapping (DTW), were used for classification between fall and non-fall events. Then, the accuracy of all the results of all the algorithms were computed and better one was implemented.

3.3.1 Algorithm 1: Threshold Based

Figure 4 shows the flowchart for the threshold-based algorithm for fall detection.

The total sum acceleration vector (Acc), which contains both static and dynamic acceleration components, is calculated from sampled data using Eq. 1.

$$\mathrm{Acc} = \sqrt{(A_x)^2 + (A_y)^2 + (A_z)^2} \tag{1}$$

where A_x, A_y, and A_z, are the accelerations (g) in the x, y, z directions.

Similarly, angular velocity is calculated from sampled data using Eq. 2.

$$\omega = \sqrt{(\omega_x)^2 + (\omega_y)^2 + (\omega_z)^2} \tag{2}$$

where ω_x, ω_y, and ω_z, are angular velocities in x, y, z directions.

When stationary, the acceleration magnitude (Acc) from triaxial accelerometer is constant (+1 g), and angular velocity is 0°/s. When the subject falls, the acceleration rapidly changes and a sudden spike is observed and the angular velocity also changes, thereby producing signals with different patterns which can be easily differentiated as fall and non-fall events. Other than the change in acceleration, the Euler angles are being used to find whether the subject has fallen or not, thus giving more precise results. Euler angles (here pitch and roll have been used) are obtained by fusing accelerometer and gyrometer data efficiently using Madgwick orientation filter.

In this algorithm, both lower fall threshold for acceleration (LFTa) and upper fall threshold for acceleration (UFTa) in combination with the upper fall threshold for gyrometer (UFTg) for fall detection are being used for better accuracy. All the threshold values were decided by trial and error method through a collection of experimental data.

First, the acceleration at that instant is computed and compared to the LFTa. If the Acc falls below the already determined LFT a threshold, data from the next

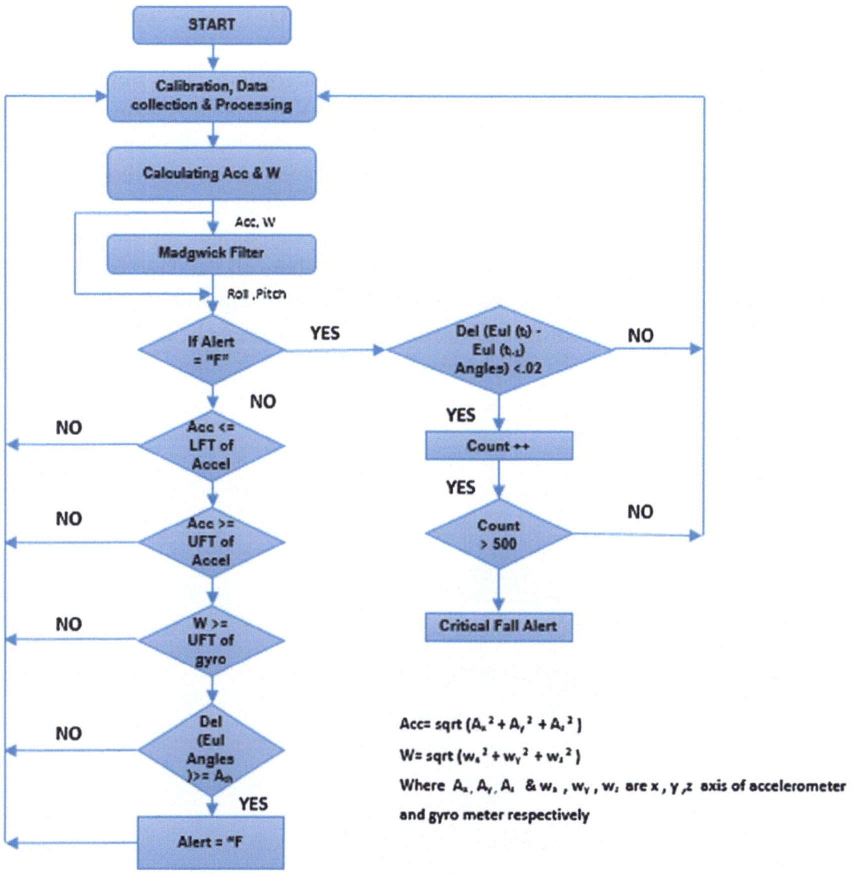

Fig. 4 Flowchart for threshold-based algorithm

0.5 s period is referred to as fall window, and all the Acc values are compared to the UFT for both the acceleration and angular velocity vector (gyrometer) to the threshold value, and if that conditions are also satisfied, then change in Euler angles is compared to Del(E) (change in Euler angles). If all the conditions are proved to be right, then "FALL ALERT" ("F") will be detected and shown.

There can be a condition where a subject has fallen and got up by his own, and it was not a critical situation. In order to distinguish between that, another condition is being introduced. If the integration of the output value is constant for a period of time (here 500 ms), then that person is not able to get up and he is in need of help. So a "CRITICAL FALL ALERT" ("C") will be detected and shown.

3.3.2 Algorithm 2: Support Vector Machine (SVM)

Figure 5 shows the flowchart of the SVM-based algorithm. The output of the Madgwick filter is given to the SVM classifier. The SVM is already trained with the training data. The feature vectors used were of mean, standard deviation, variance of all the four outputs of Madgwick filter, accelerometer (Acc), gyroscope (Gyro), roll, and pitch angles. Thus, a 12×1 vector will be fed to the SVM.

When the accelerometer buffer, gyroscope buffer, pitch buffer, and roll buffer are full, it will be sent to the feature extraction function where all the three features will be extracted for each element and these features will be compared with the already trained SVM classifier. According to the already-specified labels, if the output is a positive value (>0), a fall alert will be given, and else if it is a non-fall, the process continues.

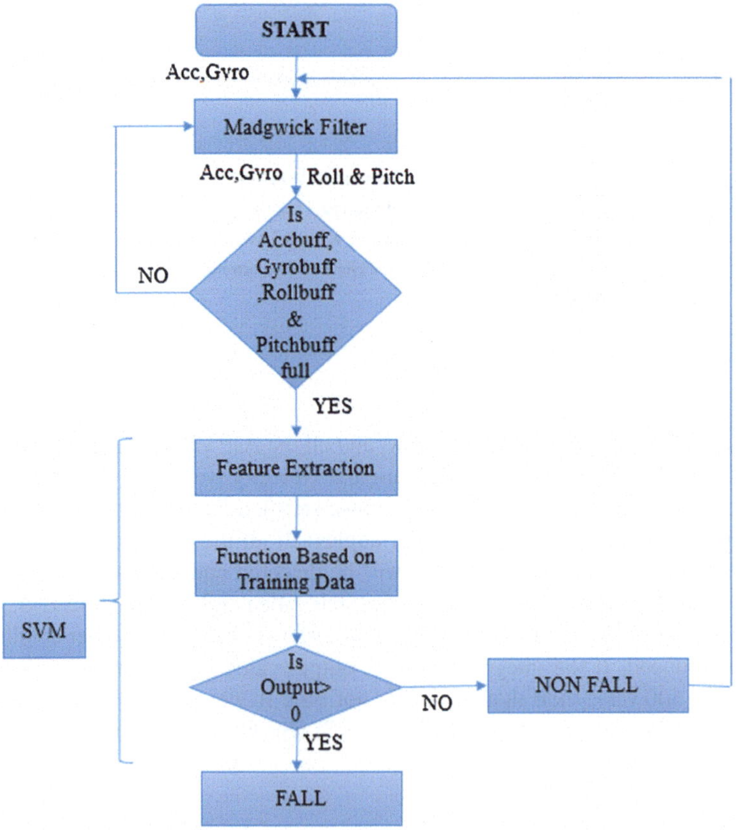

Fig. 5 Flowchart for SVM-based algorithm

3.3.3 Algorithm 3: *K*-Nearest Neighbor (*K*-NN)

Figure 6 shows the flowchart of the *K*-NN-based algorithm. The Euclidian distance (Eq. 3) is calculated for the unknown feature vector with every training feature vectors. Then, it is sorted in ascending order along with the corresponding labels, and first, *K* (=3) sorted elements are captured, and decision is made based on the labels as fall or non-fall.

$$\text{Euclidean distance} = \sqrt{(x_i - y_i)^2} \tag{3}$$

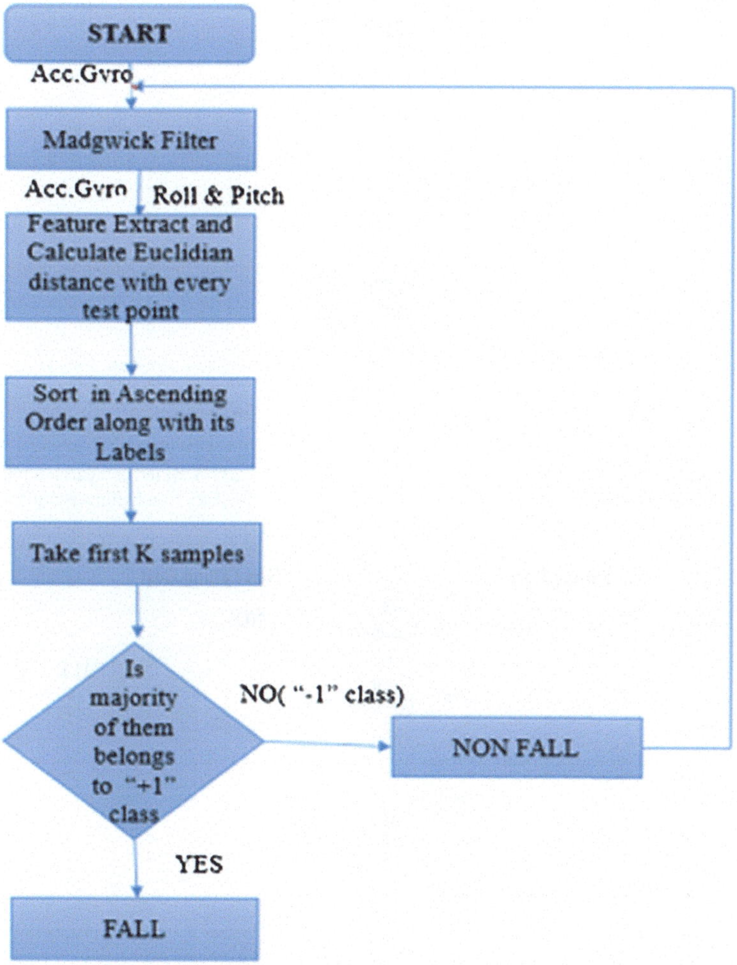

Fig. 6 Flowchart for *K*-NN-based algorithm

3.3.4 Algorithm 4: Dynamic Time Wrapping (DTW)

DTW is an algorithm for measuring similarity between two time domain sequences. Here, accelerometer pattern of the human activity is compared against accelerometer pattern of falling and non-falling pattern which is already saved. Cost of the alignment or wrapping curve is then calculated. Decision is made in favor of the combination of having the lower cost as fall or non-fall. Figure 7 shows the algorithm for fall detection using DTW.

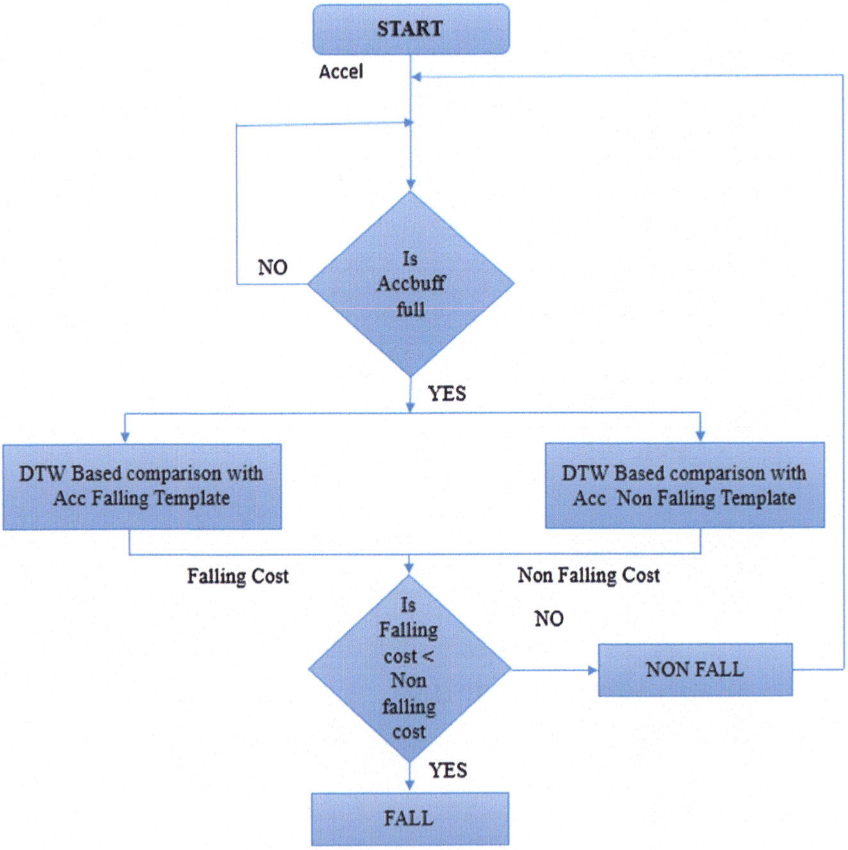

Fig. 7 Flowchart for DTW-based algorithm

4 Experimental Setup

4.1 Datasets

The study involved six volunteers (three males and three females) within the age group of 24–30 years. All the scenarios were repeated by the subjects, and 250 datasets were captured which is further used as testing and learning data.

Different scenarios were taken care while recording the datasets (Table 1). Different types of fall which includes seven activities: front fall, back fall, left side fall, right side fall, falling from cot, falling from chair and falling from stairs are being recorded. The different non-fall activities which are the normal daily activities like walking, jogging, running, sitting, bending, picking something from floor, leaning toward wall, squatting, climbing stairs up and down, trembling, slipping, sleeping on floor, and some random activities were also recorded.

The device is placed in the center part of the subject's waist to get proper datasets as shown in Fig. 8. Subjects were made to fall on bed and mattresses for safety purpose, and some sequence of images showing fall and non-fall events by the subjects is shown in Fig. 9.

Figure 9a describes a falling front event by the subject, and Fig. 9b represents a non-fall event, bending and picking an object from the ground. Proper precautions were given while taking the datasets. The subject is made to fall on bed to avoid any injuries.

Table 1 Falling and non-falling events

Falling events		Non falling events	
1	Falling front	8	Walking
2	Falling back	9	Jogging
3	Falling to left side	10	Running
4	Falling to right side	11	Sitting
5	Falling from chair	12	Bending
6	Falling from cot	13	Picking something from floor
7	Falling from stairs	14	Leaning towards wall
		15	Squatting
		16	Climbing stairs up and down
		17	Trembling
		18	Slipping
		19	Sleeping on floor
		20	Random activities

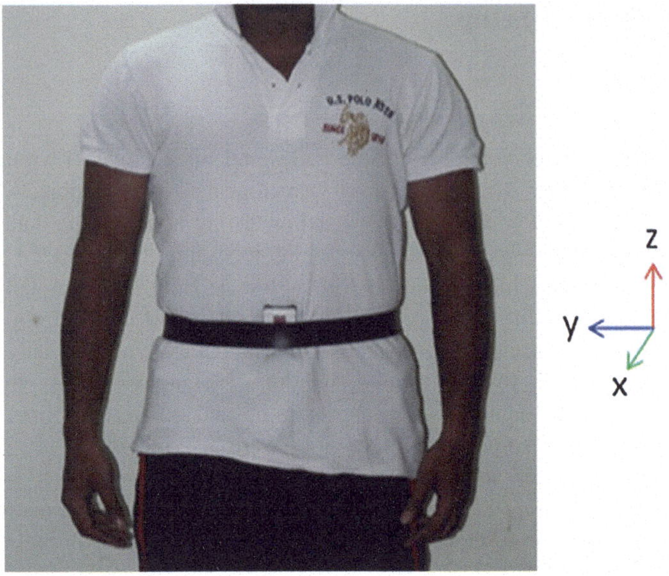

Fig. 8 Position of the device

5 Results and Discussions

5.1 Sensor Data

All the activities will have different distinct patterns as shown in Figs. 10 and 11. The simulation was done using MATLAB.

Figure 10 shows a non-fall pattern, i.e., walking. Since it is a uniform process, there is no spike. Figure 11 shows the pattern for front fall. The sudden spike is due to the subject's fall, and the upper and lower limit of these values will be compared with the already set value in the threshold-based algorithm (Algorithm 1).

5.1.1 Sensitivity, Specificity, and Accuracy of Algorithms

The sensitivity, specificity, and accuracy of all the algorithms were calculated using Eqs. 4–6 for a dataset of 100 subjects.

$$\text{Sensitivity } (\%) = \left(\frac{\text{TP}}{(\text{TP} + \text{FN})}\right) \times 100 \tag{4}$$

$$\text{Specificity } (\%) = \left(\frac{\text{TN}}{(\text{TN} + \text{FP})}\right) \times 100 \tag{5}$$

(a)

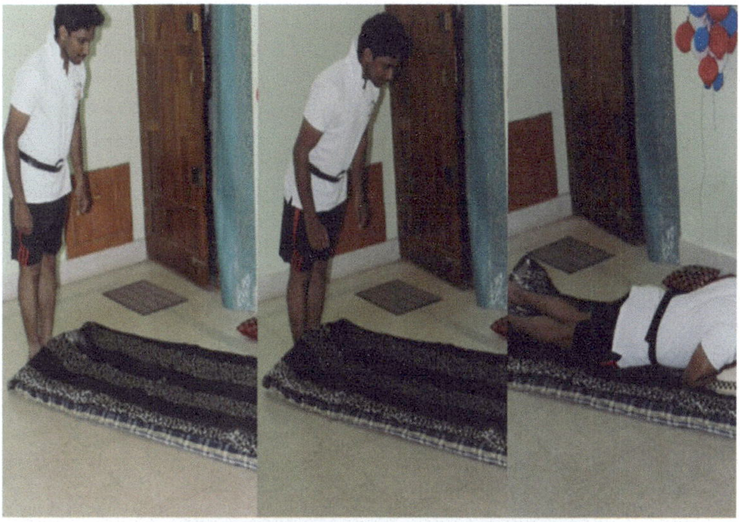

(b)

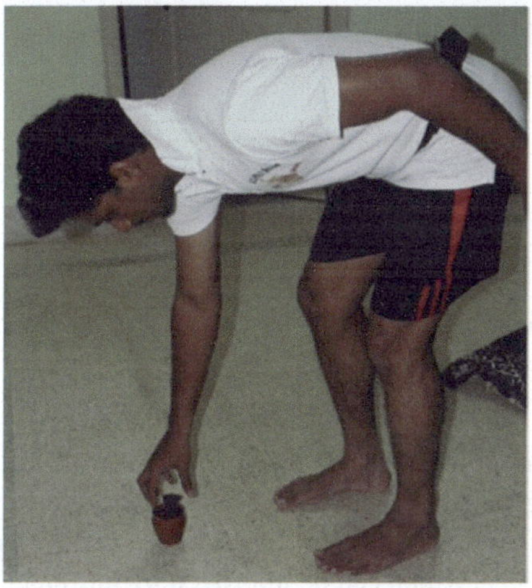

Fig. 9 **a** Sequence of images for the activities (front fall), **b** sequence of images for the activities (picking object from ground)

$$\text{Accuracy (\%)} = \left(\frac{\text{TP} + \text{TN}}{(\text{TP} + \text{TN} + \text{FP} + \text{FN})}\right) \times 100 \quad (6)$$

where

SVM Support Vector Machine
*K*NN *K*-Nearest Neighbor
DTW Dynamic Time Warping
FP False Positive

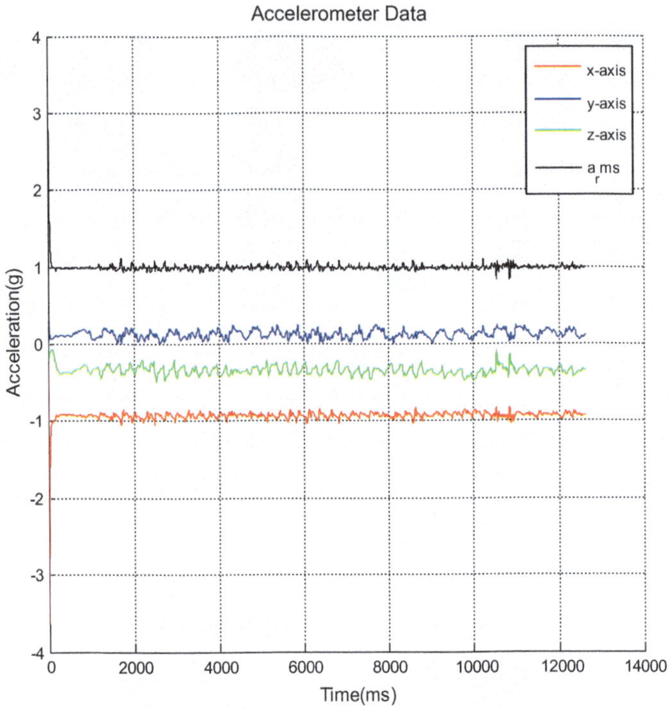

Fig. 10 Accelerometer data pattern for walking

FN False Negative
TP True Positive
TN True Negative

All units are in numbers.

Tables 2 and 3 give the sensitivity, specificity, and accuracy. From the values obtained, the SVM is found to be more accurate technique than all other techniques with an accuracy of 93% which can be improved by introducing more learning data.

5.1.2 Receiver

The device and the receiver mobile should be connected to the same Wi-fi network for receiving the result according to the prototype which was made for testing. Android-free application UDP Sender/Receiver was used for demonstrating the result. If it is normal, daily activities "A" will be shown as an indication, and if it is fall "F" will be shown and if it is critical fall; i.e., person is not able to get up for some time, "C" will be shown in the application screen. The screenshot of this demonstration in real time is shown in Fig. 12. This application was used for demonstration purpose.

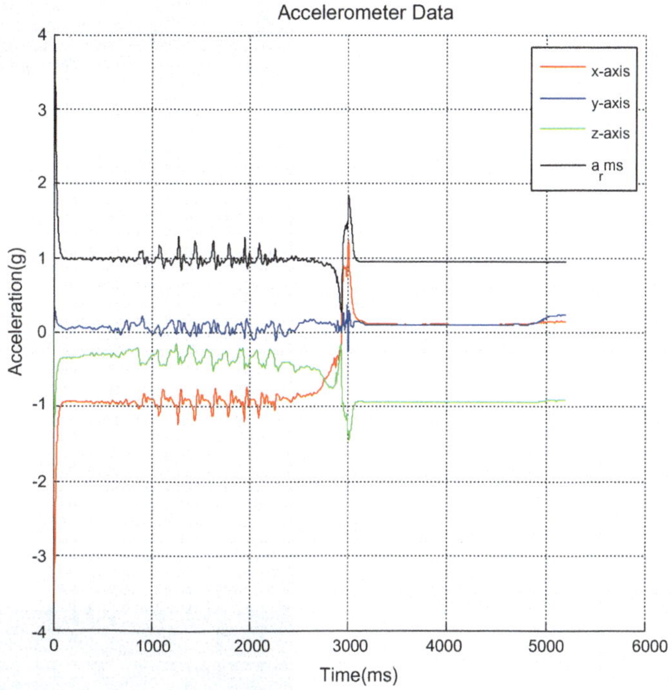

Fig. 11 Accelerometer data pattern for falling front

Table 2 Comparison of the different algorithms

Algorithms	TP	TN	FP	FN
Threshold based	58	31	4	7
SVM	65	28	6	1
K-NN	54	35	5	6
DTW	60	23	13	4

Table 3 Comparison of sensitivity, specificity, and accuracy of the algorithms

Algorithms	Sensitivity (%)	Specificity (%)	Accuracy (%)
Threshold based	89.2	88.57	89
SVM	98.49	82.26	93
K-NN	90	87.5	89
DTW	93.75	63.88	83

Fig. 12 Output obtained in UDP sender/receiver app

6 Conclusions

A prototype of a wearable wireless device for fall detection has been realized and presented in this paper; to get better accuracy, both accelerometer and gyroscope sensors are used. BMI160 has both the sensors embedded in it. The use of Madgwick filter for sensor fusion gives better orientation of the subjects and thus better results. The experiments were conducted using all the four algorithms, and a comparative study has been done. The classification based on SVM was found to be having maximum accuracy and sensitivity. The result based on this classification was sent to the UDP Sender/Receiver application for demonstration purpose. A simple application can be developed in future for displaying the result. The message can be send as an SMS along with GPS location belongs to the future scope of this project.

Acknowledgements This work was supported by Robert Bosch Engineering and Business Solutions Private Limited, Bangalore. The authors would like to thank the department colleagues, faculty, and friends who supported the work.

References

1. O'Neill T, Varlow J, Silman A, Reeve J, Reid D, Todd C, Woolf A (1994) Age and sex influences on fall characteristics. Ann Rheum Dis, 773–775
2. Kannus P, Sievanen H, Palvanen M, Jarvinen T, Parkkari J (2005) Prevention of falls and consequent injuries in elderly people. Lancet, 1885–1893
3. Holmberg AH, Johnell O, Nilsson PM, Nilsson J, Berglund G, Akesson K (2006) Risk factors for fragility fracture in middle age, a prospective population-based study of 33,000 men and women. Osteoporos, 1065–1077
4. Huynh QT, Nguyen UD, Tran SV, Nabili A, Tran BQ (2013) Fall detection system using combination accelerometer and gyroscope. In: International conference on advances in electronic devices and circuits, pp 52–56
5. Ariani A, Redmond SJ, Chang D, Lovell NH (2010) Software simulation of unobtrusive falls detection at night-time using passive infrared and pressure mat sensors. In: 32nd annual international conference of the IEEE (EMBS), pp 2115–2118
6. Wang Y, Bai X-Y (2013) Research of fall detection and alarm applications for the elderly. In: International conference on mechatronic sciences, electric engineering and computer (MEC), pp 615–619
7. Zhang Z, Kapoor U, Narayanan M, Lovell NH, Redmond SJ (2011) Design of an unobtrusive wireless sensor network for night time falls detection. In: 33rd annual international conference of the IEEE EMBS, pp 5275–5278
8. Sudarshan BG, Hegde R, Prasanna Kumar SC, Satyanarayana BS (2013) Design and development of fall detector using fall acceleration. IJRET: Int J Res Eng Technol 2321–7308
9. Kailas A (2012) Basic human motion tracking using a pair of gyro and accelerometer MEMS devices. In: IEEE 14th international conference on e-health networking, applications and services instrumentation and measurement, pp 298–302
10. Pierleoni P, Belli A, Palma L, Pellegrini M, Pernini L, Valenti S (2015) A high reliability wearable device for elderly fall detection. IEEE Sens J, 1530–1539

11. Alam F, ZhaiHe Z, Jia H (2015) A comparative analysis of orientation estimation filters using MEMS based IMU. In: 2nd international conference on adaptive science & technology, pp 86–91
12. Wu F, Zhao H, Zhao Y, Zhong H (2015) Development of a wearable-sensor-based fall detection system. Int J Telemed Appl, 576364–576374
13. Tong L, Song Q, Ge Y, Liu M (2013) HMM-based human fall detection and prediction method using tri-axial accelerometer. IEEE Sens J, 1849–1856
14. Honglun H, Meimei H, Minghui W (2013) Sensor-based wireless wearable systems for healthcare and falls monitoring. Int J, 2200–2216
15. Özdemir AT, Barshan B (2014) Detecting falls with wearable sensors using machine learning techniques. Sens J, 10691–10708
16. Dinh A, Teng D, Chen L, Ko SB, Shi Y, Basran J, Del Bello-Hass V (2008) Data acquisition system using six degree-of-freedom inertia sensor and ZigBee wireless link for fall detection and prevention. In: 30th annual international IEEE EMBS conference, pp 2353–2356
17. Mendulkar A, Kale R, Agrawal A (2014) A survey on efficient human fall detection system. Int J Sci Technol Res, 2277–8616
18. Yin X, Shen W, Samarabandu J, Wang X (2015) Human activity detection based on multiple smart phone sensors and machine learning algorithms. In: IEEE 19th international conference on computer supported cooperative work in design (CSCWD), pp 582–587
19. Liu S-H, Cheng W-C (2012) Fall detection with the support vector machine during scripted and continuous unscripted activities. Sens J, 12301–12316
20. Jantaraprim P, Phukpattaranont P, Limsakul C, Wongkittisuksa B (2012) Fall detection for the elderly using a support vector machine. Int J Soft Comput Eng (IJSCE), 484–490
21. Liu CL, Lee CH, Lin P-M (2010) A fall detection system using K-nearest neighbor classifier (Elsevier), pp 7174–7181
22. Shi G, Chan CS, Li WJ, Leung K-S, Zou Y, Jin Y (2009) Mobile human airbag system for fall protection using MEMS sensors and embedded SVM classifier. IEEE Sens J, 495–503
23. Chen J, Kwong K, Chang D, Luk J, Bajcsy R (2005) Wearable sensors for reliable fall detection. In: 27th IEEE annual conference on engineering in medicine and biology, pp 3551–3554
24. Sengto A, Leauhatong T (2012) Human falling detection algorithm using back propagation neural network. In: Biomedical engineering international conference (BMEiCON), pp 978–982
25. Hsu C-W, Chang C-C, Lin C-J (2003) A practical guide to support vector classification. Department of Computer Science, National Taiwan University, Taipei 106, Taiwan, pp 1–16
26. Sree Madhubala J, Umamakeswari A (2015) A vision based fall detection system for elderly people. Ind J Sci Technol 8(S9):167–175
27. Grace Kanmani Prince P, Hemamalini R, Immanuel Rajkumar R (2014) LabVIEW based abnormal muscular movement and fall detection using MEMS accelerometer during the occurrence of seizure. Ind J Sci Technol 7(10):1625–1631
28. Madgwick S (2010) An efficient orientation filter for inertial and inertial/magnetic sensor arrays. Computer Science and Engineering, University of Washington, pp 1–32

Mathematical Analysis of Adaptive Queue Length-Based Traffic Signal Control

Shaik Khaja Mohiddin, C. Prasanth, Gajendra Singh Rathore and C. Hemanth

Abstract *Objectives* Traffic signals are pre-timed/fixed cycle in which the duration of green light and red light is fixed irrespective of the number of vehicles at the traffic signal, which causes congestion in most cases. Most of the research done on pre-timed traffic signal focuses on deriving formulas for the better estimation of the delay experienced by vehicles. Proposed method of adaptive traffic signals can be used to minimize the congestion problem occurring at the intersection due to pre-timed traffic signal, thereby minimizing the delay to a greater extent. *Methods/Statistical Analysis* In this paper, the vehicles are considered to arrive based on a Poisson distribution. The probability generating function of queue lengths at the beginning of kth and $k + 1$th cycle is derived. Based on the queue length for each direction, the optimal duration of red and green light is derived. *Findings* In the proposed work, it is showed that the performance of adaptive queue length-based switching is efficient than the pre-timed/fixed traffic signaling. Based on the traffic arrival pattern, the threshold of the red and green light can be set to minimize the delay/congestion. *Application/Improvements* In this proposed work, the congestion problem is minimized by using the adaptive signal switching and the performance of system can be improved in heavy traffic scenario.

Keywords Adaptive traffic signal control · Delay · Fixed traffic light

---

S. K. Mohiddin · C. Prasanth · G. S. Rathore · C. Hemanth (✉)
School of Electronics Engineering (SENSE), VIT University, Chennai Campus, Chennai, Tamil Nadu, India
e-mail: hemanth.c@vit.ac.in

S. K. Mohiddin
e-mail: skhajamohiddin4@gmail.com

C. Prasanth
e-mail: mcprasanth93@gmail.com

G. S. Rathore
e-mail: gajendra0910@gmail.com

A. M. Zungeru et al. (eds.), *Wireless Communication Networks and Internet of Things*, Lecture Notes in Electrical Engineering 493,
https://doi.org/10.1007/978-981-10-8663-2_24

1 Introduction

Congestion at the traffic signal junction in city is very serious problem as density of vehicles is increasing day by day. In fixed-cycle traffic light, the duration of green light and red light is set to some default value which will not alter or change after the installation of the system. Pre-timed duration of traffic signal is based on the report in which the load information is specified on which basis the duration of the light will decide, but a number of vehicles are increasing fixed-cycle traffic light signals which are unable to cope up with the present scenario which results in congestion in most cases. To resolve the issue of congestion, dynamic control of traffic signal is proposed here. In this proposed work, the traffic signal can change dynamically with respect to the present scenario at the traffic signal and it is simulated using MATLAB. Mathematical analysis of proposed technique is also carried with this to provide theoretical support to proposed technique.

Adaptive switching of traffic signal on the basis of the queue length of the vehicle is one of the significant components of intelligent transportation system (ITS). A method [1] is used to detect the vehicle using wireless sensor network technology, in which vehicle can be monitored dynamically to overcome the disadvantages of conventional vehicle detection techniques of traffic control system in which vehicle can only detect in a fixed position. Some advantages of proposed algorithm for traffic signal control over conventional algorithm like, it eliminate the phase time when no vehicle passing across and the entire waiting vehicle pass if possible which reduces the waiting time for the single intersection. Green Light District [2] (GLD) simulator is used to generate the graph of scenario which shows effective traffic control in a real traffic road system, and algorithm is proposed which is adaptive to traffic flow to reduce the average waiting time of vehicle at traffic signals.

A method to address traffic congestion problem was proposed by using virtual traffic light (VTL) [3]. When vehicles are approaching the same intersection, they pick one of them as the leader for the intersection. The leader will act as temporary traffic light infrastructure, and then traffic light information will be created and announced by it. VANET-based algorithms [1] were used to collect and aggregate real-time speed and position information on individual vehicles to optimize signal control at traffic intersections. An online algorithm, referred to as oldest job first (OJF) to minimize the delay across the intersection, here they first grouped the vehicular traffic into platoons, and they apply the OAF algorithm, i.e., the oldest arrival first (OAF) algorithm. Mathematical model for real-time queue estimation is using connected vehicles (CV) technology [2] from wireless sensor networks. This model can be applied without signal timing, traffic volume, or queue characteristics as basic inputs. This model developed in such a way that it can be worked for both fixed-time signal and actuated signals. Here, a discrete wavelet transform (DWT) is applied to the queue estimation. The purpose of DWT is to enhance the proposed queue estimation to be more accurate and consistent regardless of the randomness in the penetration ratio. The probe vehicle is equipped with GPS and wireless communication devices from there traffic data was collected. Two types of traffic signal junctions [4], one will be simple four-way junction and each way contains two lanes. The other one is complex traffic signal which has more than four ways. To minimize unwanted delay, two algorithms proposed: one of these is earliest deadline first (EDF) and the other is fixed priority (FP) algorithm. These algorithms work efficiently in complex road scenario and heavy traffic load. For collection of information, sensors are deployed on each lane of the road which sent information to the controller via roadside unit using suitable wireless technology. Python controller is used for the proposed algorithm in this work along with SUMO which uses TCP/IP protocol, SUMO acts as a server and python acts as client. For determination of arrival rate of vehicle, Poisson distribution is used and for delay analysis [5], some assumptions like discrete-time assumption and FCTL assumptions are made. In FCTL [6] assumption, if the vehicles arrive during the residual green period can cross the intersection without delay and therefore the discharge rate of these vehicles is much higher than the discharge rate of delayed vehicles.

2 Proposed Work

2.1 Adaptive Queue Length-Based Traffic Light Control

In this paper, we proposed that congestion problems at the intersection can be minimized using the proposed method in which traffic signals can adaptively switch traffic lights according to the present scenario. In pre-timed traffic signal, the effective duration of green and red light is divided into equal time interval, due to which congestion takes place in heavy traffic scenario at intersection. Delay experienced by vehicle in pre-timed signal is negligible for short length queue but as the queue length increases, delay experienced by vehicle also increases.

In this proposed work, MATLAB software tool is used for the coding of the scenario, in which the arrival rate of the vehicle is varying continuously after every slot on each lane of road. Different arrival time of vehicle will cause congestion in fixed-cycle traffic light. Therefore, using MATLAB code proposed technique which is used for switching of traffic light according to the number of vehicles arrived at the junction. In this technique, red light and green light duration will increase or decrease as per the number of vehicles arrived at the junction. So that proposed

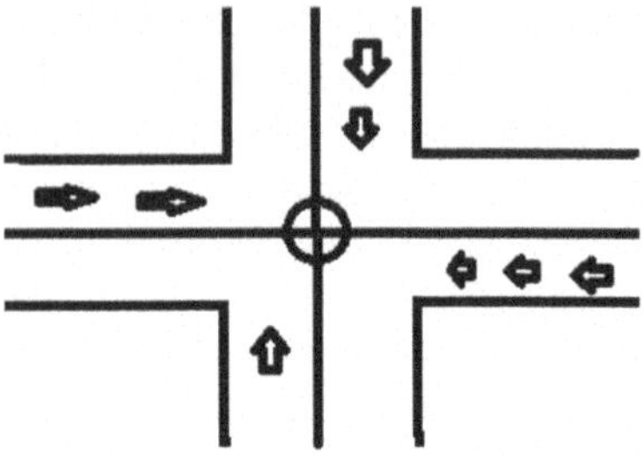

Fig. 1 Scenario for proposed method

technique will work on the scenario which is shown in Fig. 1. If the number of vehicles same in both directions, then priority is given for vehicles which arrived from longest lane.

2.2 *Mathematical Analysis*

For the mathematical analysis of the proposed technique, following assumptions are made:

Assumption 1. (Discrete-time assumption) The time axis is divided into the interval of unit length, so called slot of green (g) and red (r) light duration. Therefore, cycle time is

$$C = g + r \tag{1}$$

Assumption 2. (Independence assumption) Let $A_{k,n}$ denotes number of vehicles that arrive at the intersection during slot k in cycle n, which are independent and identically distributed.

Assumption 3. To measure the arrival rate of vehicle, Poisson distribution is used and instantaneous decision is taken.

In this work, based on the basic assumptions, mathematical equations can be derived using which the duration of green light will alter/changer exactly according to the number of vehicles present at the intersection. In this case, the delay experienced by the vehicle will be less than the delay experienced at the fixed-cycle traffic signal and also provide the theoretical support to the proposed technique. Following equation will be derived for the switching of the green and red light, respectively, as shown in Fig. 2 with the help of basic assumption.

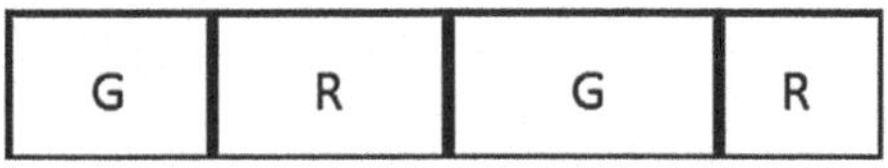

Fig. 2 Adaptive traffic cycle duration

For the determination of arrival rate of vehicle, Poisson distribution is used. Q_{Gn}^{k} is the initial queue length at the intersection, where n will be lane identifier for each lane. For the switching of the signal first threshold value, G_t will be set which is the maximum number of vehicle can be served in one cycle. Let A_{k+1} be the new arrival at the junction before/after the completion of one cycle. α and β are the numbers of served vehicle in one cycle.

For Path 1 (p1):

$$Q_{G1}^{k+1} = \begin{cases} Q_{G1}^{k} + A_{k+1} - \alpha_1; & Q_{G1}^{k} > 0 \\ A_{k+1}; & Q_{G1}^{k} = 0 \end{cases} \quad (2)$$

where, $\alpha_1 = \min\{Q_{G1}^{k}, G_t\}$.

And G_t = Threshold value of Green interval.

For Path 2 (p2):

$$Q_{G2}^{k+1} = \begin{cases} Q_{G2}^{k} + A_{k+1} - \alpha_2; & Q_{G2}^{k} > 0 \\ A_{k+1}; & Q_{G2}^{k} = 0 \end{cases} \quad (3)$$

where $\alpha_2 = \min\{Q_{G2}^{k}, G_t\}$.

For Path 3 (p3):

$$Q_{G3}^{k+1} = \begin{cases} Q_{G3}^{k} + A_{k+1} - \beta_1; & Q_{G3}^{k} > 0 \\ A_{k+1}; & Q_{G3}^{k} = 0 \end{cases} \quad (4)$$

where $\beta_1 = \min\{Q_{G3}^{k}, G_t\}$.

For Path 4 (p4):

$$Q_{G4}^{k+1} = \begin{cases} Q_{G4}^{k} + A_{k+1} - \beta_2; & Q_{G4}^{k} > 0 \\ A_{k+1}; & Q_{G3}^{k} = 0 \end{cases} \quad (5)$$

where $\beta_2 = \min\{Q_{G4}^{k}, G_t\}$.

From the above equations, we can evaluate the arrived vehicles at the intersection and for the calculation of green period, we have to find α, β and γ. From path 1 and path 2, we can get

$$\alpha = \mathrm{Max}\{\alpha_1, \alpha_2\} \quad (6)$$

Similarly, from path 3 and path 4, we get

$$\beta = \max\{\beta_1, \beta_2\} \quad (7)$$

Then, we get

$$\gamma = \max\{\alpha, \beta\} \tag{8}$$

where γ is final duration of green light for vehicle to be served and the red period will be the minimum of α and β. Now from the probability theory, we are calculating pdf at random queue length as j.

Probability for p1 at queue length j and $x > \alpha_1$,

$$P = \left\{Q_{G_1}^{k+1} = j\right\} = \sum_{x=\alpha_1+1}^{j+\alpha_1} P\left\{Q_{G_1}^{k} = x\right\} P\{A_{k+1} = j - x + \alpha_1\} \forall\ j \geq 1 \tag{9}$$

Probability for p1 at queue length j and $x < \alpha_1$

$$P = \left\{Q_{G_1}^{k+1} = j\right\} = \sum_{x=0}^{\alpha_1-1} P\left\{Q_{G_1}^{k} = x\right\} P\{A_{k+1} = j\} \tag{10}$$

Probability for p1 at queue length j and $x \leq \alpha_1$

$$P = \left\{Q_{G_1}^{k+1} = 0\right\} = \sum_{x=0}^{\alpha_1-1} P\left\{Q_{G_1}^{k} = x\right\} P\{A_{k+1} = 0\} \tag{11}$$

From Eq. (11), we can say that the probability at initial queue length ($P\{Q_{G_1}^{k} = x\}$) is dependent on initial vehicles queue length, and here the number of arriving vehicles is zero ($P\{A_{k+1} = 0\}$).

3 Results and Discussion

We generate random packets in MATLAB using exponential distribution in four lanes, and we fix a threshold for green duration and write a code for our mathematical analysis. Figures 3 and 4 show the arrival rate of vehicle at the junction, threshold value of the green light duration, and data after completion of each cycle. Figure 5 shows the number of vehicles arriving on each lane, and the length of the vehicle is assumed to be ten units which take 1 s to cross the intersection. Figure 6 shows the number of vehicles per particular slot with respect to green and red duration by using the data available in the result section, the traffic lights can be dynamically changed and duration of the green light and red light will adaptively change according to the arrival rate of the vehicles.

Workspace

Name	Value	Min	Max
A	50	50	50
ArrivalTime	[10077,10022,10014,1...	10014	10154
AverageArrivalTime	[100,100,100,100]	100	100
AveragePacketLen...	50	50	50
B	50	50	50
Buffer_Threshold	1000	1000	1000
C	50	50	50
CurBufferSize	[910,880,950,840]	840	950
D	50	50	50
G	[50,50,50,50]	50	50
M	4	4	4
PacketBuff	<4x500 double>	0	95
PacketLength	[10,10,10,10]	10	10
SlotTime	2	2	2
TotalTime	10000	10000	10000
TotalTimeInSecon...	20000	20000	20000
U	45	45	45
V	34	34	34
X	41	41	41
Y	38	38	38
a	91	91	91
alpha	50	50	50
alpha2	41	41	41
b	88	88	88
beta	50	50	50
beta2	45	45	45
c	95	95	95
d	84	84	84

Fig. 3 Output data of mathematical analysis-I

Workspace

Name ▲	Value	Min	Max
SlotTime	2	2	2
TotalTime	10000	10000	10000
TotalTimeInSecon...	20000	20000	20000
U	45	45	45
V	34	34	34
X	41	41	41
Y	38	38	38
a	91	91	91
alpha	50	50	50
alpha2	41	41	41
b	88	88	88
beta	50	50	50
beta2	45	45	45
c	95	95	95
d	84	84	84
green1	50	50	50
green2	45	45	45
i	4	4	4
j	4	4	4
k	4	4	4
n	4	4	4
red1	50	50	50
red2	41	41	41
t	10000	10000	10000
threshold	50	50	50
u	[45,45,45,45]	45	45
v	[34,34,34,34]	34	34
x	[41,41,41,41]	41	41
y	[38,38,38,38]	38	38

Fig. 4 Output data of mathematical analysis-II

Variables - PacketBuff

G × PacketBuff ×

PacketBuff <4x500 double>

	1	2	3	4	5	6	7	8
1	97	10	10	10	10	10	10	10
2	99	10	10	10	10	10	10	10
3	14	10	10	10	10	10	10	10
4	12	10	10	10	10	10	10	10

Fig. 5 Number of vehicles arrived at junction

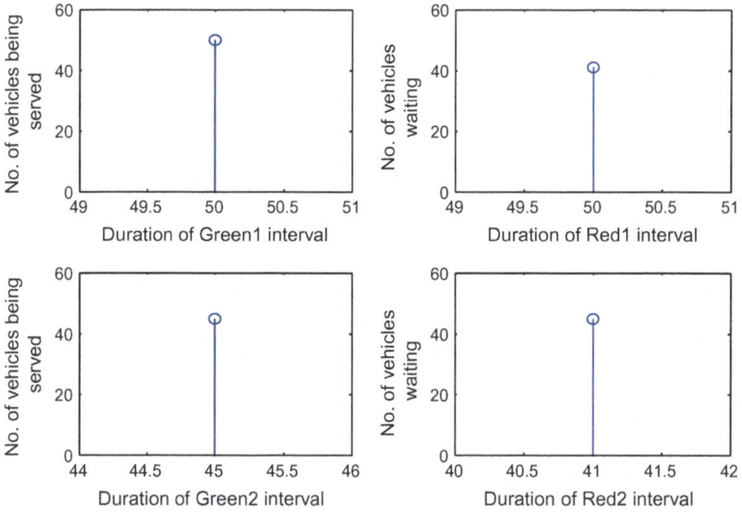

Fig. 6 Number of vehicles served per particular red and green duration

4 Conclusion

In this paper, we proposed adaptive switching of traffic light to minimize the congestion problem. The probability generating function of the queue lengths at different lanes was calculated and based on it, the duration of red and green signal is determined. The proposed scheme reduces the traffic congestion problem at major intersection and outperforms the fixed traffic light switching mechanism.

References

1. Pandit K, Ghosal D, Michael Zhang H, Chuah C-N (2013) Adaptive traffic signal control with vehicular ad hoc networks. IEEE Trans Veh Technol 62(4)
2. Tiaprasert K, Zhang Y, Wang XB, Zeng X (2015) Queue length estimation using connected vehicle technology for adaptive signal control. IEEE Trans Intell Transp Syst 16(4)
3. Shi J, Peng C, Zhu Q, Duan P, Bao Y, Xie M (2015) There is a will, there is a way—a new mechanism for traffic control based on VTL and VANET. In: IEEE 16th international symposium on high assurance systems engineering
4. Ahmad A, Arshad R, Mahmud SA, Khan GM, Al-Raweshidy HS (2014) Earliest deadline based scheduling to reduce urban traffic congestion. IEEE Trans Intell Transp Syst 15(4)
5. Wenjie C, Lifeng C, Zhanglong C (2005) Delay analysis for the fixed cycle traffic light queue. In: International conference on parallel processing workshops (ICPPW'05), TU Shiliang Department of Computer Science and Engineering, Fudan University
6. van Leeuwaarden JSH (2006) Delay analysis for the fixed-cycle traffic-light queue. Transp Sci 40(2)

Wireless Data Acquisition and Communication System for Automated Guided Vehicle

Sujay Ballal, Mohan Jagannath and K. Arun Venkatesh

Abstract The transportation of materials from/to storage and unloading points usually requires human for operating forklifts. There is a need for robotic automated guided vehicle (AGV) to minimize the human intervention and also a secured transmission system between the user and the AGV. The objective of this study is to develop wireless data acquisition and communication system for AGV. The core of the system lies in the wireless communication protocol and requires secure transmission of data to be exchanged between the user and the AGV, hence the Transmission Control Protocol/Internet Protocol (TCP/IP). The data acquisition from the AGV is accomplished with the help of sensors mounted on the vehicle that can continuously collect the incoming signals along the path, and they are transmitted wirelessly via Wi-Fi and the information is displayed to the user on a laptop. The DC to DC converter helps in charging the rechargeable battery to the nominal voltage. The warehouse management system is achieved with the MS access at the backend and the LabVIEW at the front end for creating, inserting, deleting, and updating the data wirelessly via Wi-Fi. The analog data from the sensors was collected with each sensor having a threshold level for obtaining the maximum or minimum level that is encountered during the travel. The data from the AGV is obtained to the user by establishing connection through IP address. The warehouse management system provides more accuracy and prevents human blunders when compared to traditional system. They are able to insert or update in fraction of second, and these data can be viewed by anyone. The battery management system is able to charge the rechargeable battery at a faster rate. The AGV is able to transport goods from one location to the other in a predefined path. The data acquisition from the sensors provides a means of navigation system for the

S. Ballal
Wipro Technologies, Bengaluru, Karnataka, India

M. Jagannath (✉)
School of Electronics Engineering, Vellore Institute of Technology (VIT), Chennai, Tamil Nadu, India
e-mail: jagannath.m@vit.ac.in

K. Arun Venkatesh
Yashika Industries, Chennai, Tamil Nadu, India

A. M. Zungeru et al. (eds.), *Wireless Communication Networks and Internet of Things*, Lecture Notes in Electrical Engineering 493,
https://doi.org/10.1007/978-981-10-8663-2_25

vehicle. These vehicles are mainly used in the industrial warehouse that reduces the labor cost and enhances accuracy.

Keywords Automated guided vehicle · Transmission control protocol Wi-Fi · Wireless communication

1 Introduction

A material handling equipment (MHE) has been significantly used for transporting goods from one place to another especially in an industrial warehouse. The goods are travelled without the intervention of humans. The goal is to reduce damage, protection of materials, enhance safety, and improve the working conditions. The area in which it is being used has increased significantly. The AGV is an essential tool for automation transportation, loads and unloads which can link and adjust the distinct logistic systems [1].

The advancements of robotic technologies have led the industries in adopting further of automation to increase the accuracy and production quality, and thus provides a better management of time. Traditionally, at manufacturing systems the AGVs were mostly used. Currently, AGVs are popular for transportation tasks in warehouses. The vehicle is robust and user-friendly to access it. On providing the commands to the AGV, it travels accordingly to the specified location [2]. The commands are provided through the laptop/tablet/PC anywhere in the industrial warehouse. The AGV senses the command only through the wireless communication.

This will make the work much simpler since user may not be with the vehicle all the time. A secured method of wireless communication is necessary so that the data packets are not lost in the midway [3]. The easiest and the best way for the wireless communication is the wireless fidelity (Wi-Fi).

Battery management of the AGV plays an important role in moving the autonomous vehicle. The charging circuit is very much essential for charging the rechargeable battery to the maximum level. With the DC to DC converter circuit, the AC supply from the power source is converted into the suitable form for charging the battery [4]. The obstacle detection can be detected with the ultrasonic sensor, and the tilt of the vehicle can be obtained using accelerometer and the human in the path of the AGV is detected by the pyroelectric infrared (PIR) sensor attached to the vehicle.

The user interface will enhance the security of the AGV by not allowing unauthorized user to take control over it. Only the authorized user can get access into battery management system and navigation. This makes the vehicle more secured. The raw materials and finished products require storing to provide to the customers whenever required. Storage involves legitimate plan for preserving the materials from the production stage until the requirement arises. The large-scale storage done in a specific manner is termed as 'warehousing.'

The warehouses exist to provide the details regarding the products that are currently available. The organization mainly focuses on enhancing customer service. The best way to enhance customer service without added long-term expense is implementing a warehouse management system. This system increases efficiency by providing best service.

The first AGV was introduced during 1950s. The AGV can be simply stated as that it is a 'driverless' vehicle powered by electric motor and batteries. And also it can be called as unmanned vehicle that is controlled by a computer. AGV can be compared with the forklift driven by human but the former is completely automated one.

AGVs are manufactured in different sizes and are capable of carrying different loads, from no load up to a few tons. The AGVs environment can differ from air-conditioned offices with carpet floors to hospital hallways to factories with concrete floors [5].

Last generation of AGVs incorporated wire guidance has now been replaced with the laser technology that permits to locate any obstacles and maneuvers more accurately. The AGV is equipped with wireless communication protocol that allows exchanging the data between the AGV and the computer. The computer provides the commands to the AGV. In response to the command, the AGV will acknowledge whether the command is executed successfully. The wireless communication involves wireless BiM transceiver module. The commands such as movement of the vehicle are fetched from the computer, and the AGV can either receive the command or transmit but cannot operate simultaneously [6].

The most popular wireless communication network is the wireless local area network (WLAN) which is an extension of Ethernet built on the IEEE 802.11 standard. This is commonly called as Wi-Fi. Due to the property of the shared radio channel, the errors in the channel and path loss do impact that may lead to retransmission of the data packets resulting in network delays. The carrier sense multiple access (CSMA) mechanism that is used by Wi-Fi produces delays due to the heavy traffic in the network [7].

A robotic arm is controlled using Raspberry Pi wirelessly via Wi-Fi. A smartphone is operated from a remote location for controlling the robotic arm. The delay and the server issues are resolved as the Wi-Fi is implemented that has major usage of Internet [8].

The paper [9] offers a charging method for AGV that is simpler and a practical one. The traditional approach of wire charging method is improved by a non-contact wireless charging method. The approach is based on the variable frequency control, realized in a closed-loop control manner. A lead–acid cell contains a transmitter and receiver that are not in contact but are coupled electromagnetically. The strategy that is being used in this method is the constant-current charging. This is accomplished by seizing the current and voltage and altering the frequency [9].

The vehicle is controlled wirelessly, and the user is not required to approach the AGV to enter the coordinate paths for its functioning. On the other hand, the user is allowed to provide commands from a remote site. The approach is implemented in the low-cost RF transceivers. The software was developed with PIC C compiler. The action for the AGV is provided on the basis of x, y position. The x, y coordinated paths are transmitted to the AGV with the help of wireless communication channel. The actions are such as Stop, Left, Right, and Forward. The AGV receives the commands in a two-bit data that in turn controls the AGV. The wireless module is implemented in the RF radio module. The system is composed of TLP434A transmitter and RLP434 receiver [10].

2 Methodology

In Fig. 1, the NI Single-Board RIO (NI sbRIO) controller is powered up with the rechargeable battery that has a charging circuit which enables the battery to charge to the nominal voltage. The battery level can be visually seen on laptop through TCP/IP communication via Wi-Fi to the user. The tilt of the vehicle is sensed through the accelerometer placed on the vehicle. When the vehicle tilt exceeds a certain threshold level, a warning message will be obtained to the user present in the remote location wirelessly using Wi-Fi. The presence of the human is detected with the PIR sensor.

The infrared radiation emitted from the human body is captured and sent to the controller which in turn transmits the data wirelessly via Wi-Fi. The obstacle detection ultrasonic sensor prevents the vehicle from being hit to the incoming obstacle. This data will be sent to the user present at the remote location through TCP/IP via Wi-Fi wirelessly. The user at the remote location can monitor the status of the vehicle. There is also an additional feature that is implemented for the inventory control of the warehouse. The goods that are being transported from the warehouse to the customer delivery point can be manually operated by updating the remaining quantities available within the warehouse. Also the user can keep a track on the stock.

The user can keep a track on the inventory control within the warehouse. The server can be kept anywhere in the industrial warehouse, and the user is allowed to access it wirelessly via Wi-Fi. The GUI acts as the client wherein the user can update, delete, or insert the data into the inventory. All the above processes are obtained on a user-friendly graphical interface built on the laptop with Wi-Fi enabled, designed to make the work faster and easier for the user. The user must provide credentials for getting access into it. The heart of wireless communication is the IEEE 802.11 standard. The Wi-Fi enables the user and the remote computer to communicate wirelessly. Network communication is performed by the IEEE 802.3 Ethernet standard called TCP/IP protocol. This one is the most modern technology. It mainly consists of two primary elements which communicate with each other (i) a server consisting of NI sbRIO with a wireless router and (ii) a client laptop which is Wi-Fi enabled. The client laptop sends/receives data to/from the NI sbRIO using TCP/IP packets via Wi-Fi.

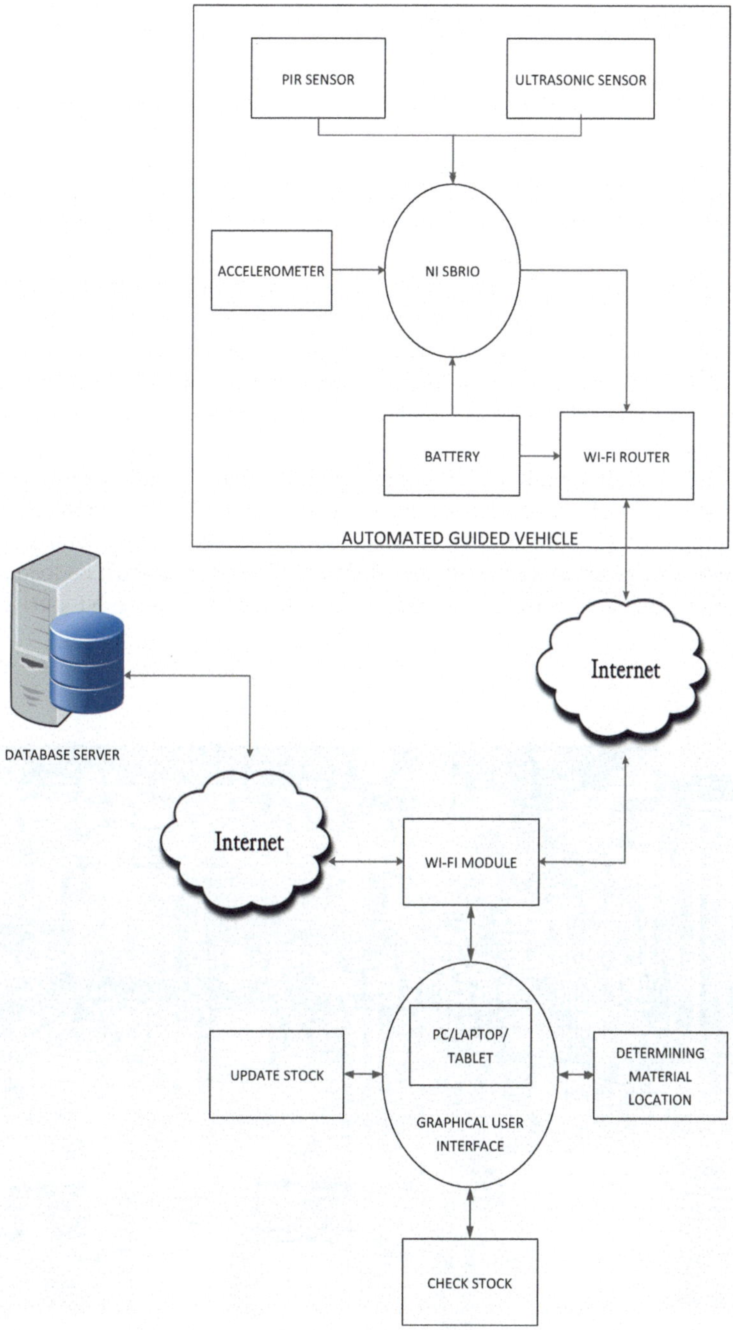

Fig. 1 Proposed AGV system design model

2.1 Hardware System Design

The hardware system design focuses on sensor modules, controller unit, communication, and battery management system.

NI sbRIO

NI sbRIO, as shown in Fig. 2, is an embedded controller with real-time processor that has the capability of data acquisition. This is a user-configurable field programmable gate array (FPGA) device that can be configured any number of times.

The NI sbRIO's GPIO port is situated on the printed circuit board on their left side. It is a 50-pin port with 16 analog inputs, 4 analog outputs and 4 digital outputs. NI sbRIO is powered up by providing a voltage of 9–30 V. The Ethernet port available in the NI sbRIO will enable to connect the wireless router for wireless communication.

With suitable configurations, the controller behaves as a stand-alone device. The various sensors such as accelerometer, ultrasonic, and PIR are interfaced to the analog inputs of NI sbRIO. The configuration of the NI sbRIO with real-time application will generate a file with an extension of .rtlexe that must be loaded into NI sbRIO to enable to act independently. The Ethernet port of the NI sbRIO must

Fig. 2 Embedded controller with real-time processor (NI sbRIO)

be connected with the wireless router since the controller does not have a built-in Wi-Fi module for wireless transfer of information.

Accelerometer

The accelerometer is a device mainly intended to detect the tilt in the vehicle. The accelerometer measures the angle at which the vehicle is tilted with respect to the ground. A more reliable method of measuring the tilt is to use the 2 axis X and Z that is capable of measuring from 90° to −90° as shown in Eqs. (1) and (2). The values will be generated when the vehicle is traveling. In order to obtain the angle at which the vehicle is being tilted, the raw data is converted into degrees.

$$\tan\theta = \frac{x}{z} \quad (1)$$

$$\theta = \arctan\left(\frac{x}{z}\right) \quad (2)$$

The predetermined threshold level is set beyond which the vehicle will topple. The degree at which the vehicle is tilted is obtained to the user at remote location through wirelessly through TCP/IP via Wi-Fi. The user will be notified about the vehicle orientation. If the tilt of the AGV exceeds the threshold level, then the rollover of the vehicle may happen which will be reported to the user wirelessly.

PIR Sensor

The detection of the human body is achieved through PIR sensor. PIR sensor is capable of generating the electrical charge when infrared radiation falls on it. It is a crystalline material with a Fresnel lens that acts a filter. The infrared signal obtained from the human body is concentrated to the element. When the human motion is detected, the sensor outputs a low-voltage value, whereas a high voltage is obtained when there is no motion of the human. This will help the vehicle to stop automatically during human presence on the path of the AGV so that the vehicle is not affected. These values will be sent to the user through TCP/IP via Wi-Fi wirelessly so that the user will be aware of the human's presence along the travel of the AGV.

Ultrasonic Sensor

The distance of an object is measured by calculating the time taken for the sound to hit the object and return from that object. Ultrasonic sensor mainly consists of two units. One unit produces the sound and the other unit catches the reflected echo from the object. The sound generated from the AGV using one unit of ultrasonic sensor triggers the timer. The other unit of the ultrasonic sensor receives the arrival of the signal after hitting the object. With the timer, AGV is able to measure the distance from the obstacle. Generally, the distance of the obstacle is half of the distance travelled by the sound from the AGV. These distances are transmitted wirelessly to the user located at the remote location through TCP/IP via Wi-Fi wirelessly.

Battery Management System

This battery management is vital for AGV since the battery is required for driving the motor that is to be travelled all along the industrial warehouse. This system does the operation of powering up many units associated with it such as NI sbRIO, sensors, and router. For charging the battery, it is important to design a circuit as shown in Fig. 3 which is capable of generating power to the battery at a faster rate.

Initially, a power supply of 230 V is applied to the bridge rectifier that converts the AC source into a DC source. The bridge rectifier's output is connected to the capacitor, which will act as filter circuit. The output from the rectifier circuit is given to high-frequency switch. The high-frequency switch is usually a fast switching power semiconductor device such as a MOSFET. This device is turned ON and OFF intermittently with the application of pulse width modulation (PWM) signal, and the unregulated DC voltage is applied to the primary of the high-frequency transformer. The switching pulses are normally fixed frequency and adjustable duty cycle. Thus, a bipolar train of voltage pulse of suitable magnitude and duty cycle appears at the transformer secondary. This bipolar voltage pulse train is rectified by an ultrafast high-speed rectifier and then smoothened by the output filter, which is a capacitor. The compensation for fluctuations (ripple) in the rectified input voltage is accomplished by the voltage feedback using fixed frequency, PWM. The duration of the on time (i.e., the duty cycle) is varied on a cycle-to-cycle basis to compensate for changes in the rectified input voltage and output load. The voltage feedback loop subtracts the DC output voltage from the

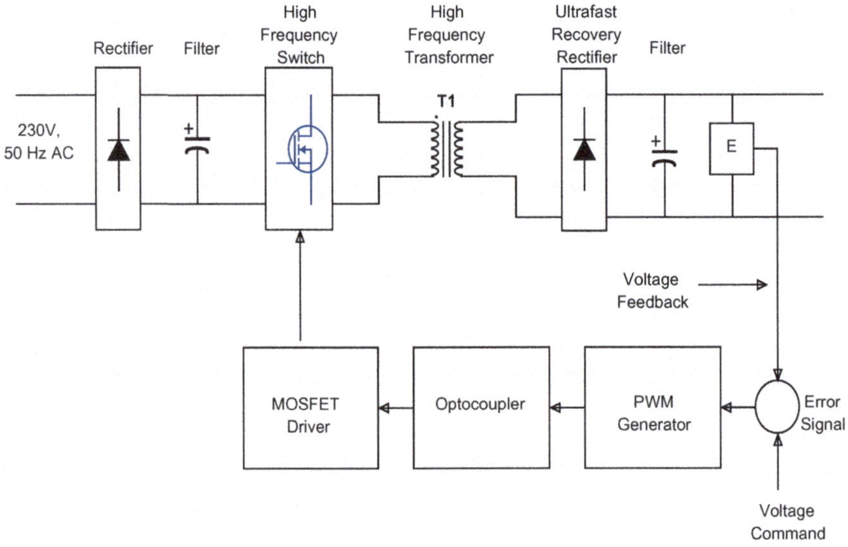

Fig. 3 A charging system for rechargeable battery

voltage command. An error signal is produced whenever the DC output voltage differs from the voltage command.

The error signal is used to adjust the duty cycle of the switching signals applied to the high-speed electronic switches so as to correct the error on the DC output voltage. For instance, a positive polarity happens when the DC output voltage is lower than the voltage command, the error signal increases the duty cycle. Therefore, the value of the RMS ac voltage at the high-frequency power transformer increases, and thus to compensate the error, the DC output voltage increases. The error signal is zero when the DC output voltage equals the voltage command and the system in equilibrium by duty cycle sets to the exact value.

With the help of the charging circuit, the battery will be charged up to maximum nominal voltage. The best rechargeable battery is the lithium-ion rechargeable battery. Voltage charging method will charge the lithium-ion battery. The charged battery will help the vehicle to maneuver inside the industrial warehouse without interruption. The monitoring of the battery enables the operator as well as AGV to charge the battery when the battery level is low. As soon the battery becomes low, the AGV will automatically move towards the charging station. The NI sbRIO with LabVIEW accomplishes this goal of monitoring the battery of AGV. The voltage and the level of the battery of the AGV will be wirelessly transmitted to the user. The user will be able to view the level of the battery from a remote location using TCP/IP via Wi-Fi.

Wireless Communication System

The way of handling the network through wirelessly is called Wi-Fi. It enables the mobile computing devices to connect to the Internet easily. The computers can be connected anywhere in home, office without requiring wire. Using radio networks, the computers are connected to the network. The antennas and routers transmit the radio signals that will be picked up by the Wi-Fi receivers. The range of the router is 100–150 ft., and the computer is able to receive the signal that can connect to the device.

A router is a device that is specialized to allow the data to move from one network to another across two or more networks. This allows the exchange of data very efficiently that is being managed in an organization. They are capable of creating a network that can exchange the data between routers on that network. The computer can connect to the Internet and communicate with another computer on that network by transferring the data to the default gateway. The computer connected to the local router has the same default gateway on that same network.

2.2 Software System Design

The software system design focuses on protocol communication, warehouse management, and graphical user interface system.

TCP/IP Communication

The shorter version of OSI model is the TCP/IP protocol. Transmission Control Protocol (TCP) the upper layer manages the assembling of messages into tiny units that can be sent over the Internet and collected by a TCP layer that reassembles the received units into the original message. The lower layer, Internet Protocol (IP) handles the address so that the message is delivered to the right destination. The computer checks each gateway on the network and forwards the message to the particular address.

TCP is a connection-oriented protocol that requires connection to be established before the message being transmitted between each other. The connection remains open between client and server until either server or client terminates the connection. The TCP/IP server program is deployed into NI sbRIO. A wireless module is necessary for communication from the NI sbRIO to the user. The best choice for the wireless module is to use a router since the Internet connection will be available in the industrial warehouse. The sensor values and the battery voltage level are acquired continuously from the NI sbRIO stand-alone real-time system that is fetched into the TCP/IP server code.

The user can connect to the controller by providing the IP address of the stand-alone system and can read those values anywhere in the industrial warehouse through wireless communication. The communication between the user and the stand-alone system occurs wirelessly via Wi-Fi.

Warehouse Management System

The warehouse management system is developed with the view to improve transportation of goods very efficiently. It enables the user to know the exact status of all the goods present in the warehouse. With this system, the user can instruct the AGV to move to the area of interest for transportation of goods. The stock in the warehouse can be handled very effectively with the option of the user to select the particular item and can get the items present in the warehouse.

The backend database is built upon Microsoft Access and the LabVIEW for the front-end interface. The user on the client side provides the command for some operations such as the table creation, inserting new data into the record set, removing the record set, updating the field, and retrieving of complete data from a remote location to server.

One of the systems acts as the server in providing the information from the client onto the laptop which reduces human effort and can be operated anywhere in the industrial warehouse. The server system has built-in Wi-Fi enabled with the laptop which can be easily communicated with the help of TCP/IP protocol. Since it is a secured way of transmission, the data from the user cannot be lost while transmission takes place.

Graphical User Interface

This is the first graphical user interface (GUI) that will appear to the user when user wishes to access the AGV. The GUI is built on the laptop using LabVIEW.

The user will be granted access to choose the desired operations only when the credentials are matching with the users that are in the local database. Once the login is successful, the user can choose either sensors or the monitoring of the battery. The user obtains these values from the AGV wirelessly via Wi-Fi through TCP/IP communication.

The user interface has also an option of choosing either change the password or add/view/delete user. The administrator has complete control over the adding a new supervisor or an operator. The supervisor is for the maintenance of AGV with all functions of the administrator but cannot add/view/delete or changes the password of the user. The operator can simply monitor the vehicle and has a close contact whether the vehicle is operating without interruption.

3 Results and Discussion

The proposed model served as a dedicated system for wireless data acquisition and communication for AGV. The recurrence control technique is carried out for achieving the entire system flow. The charging circuit designed for charging the battery yields a consistent voltage of 12 V by varying the duty cycle and applying constant frequency. The user will receive these parameters wirelessly via Wi-Fi through TCP/IP communication as shown in Fig. 4. The capacity of the battery is indicated to the user, and status will also be visible.

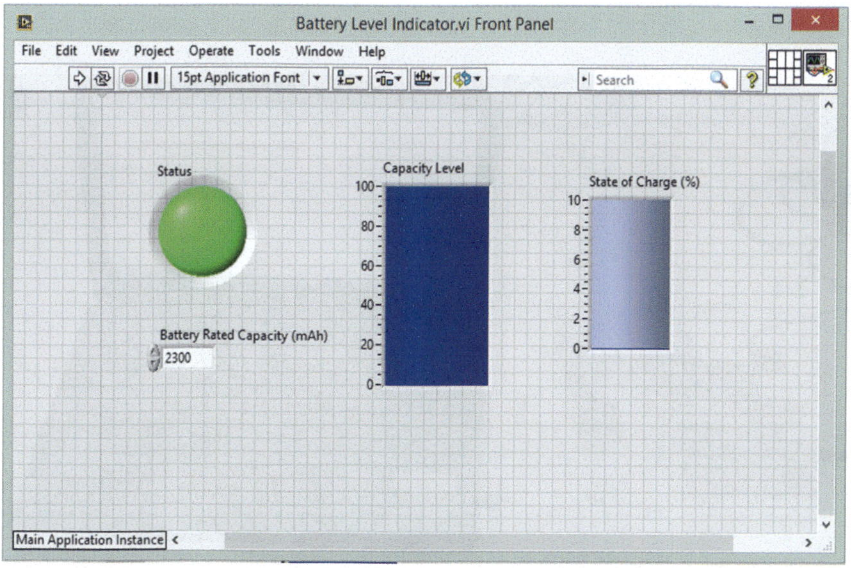

Fig. 4 Battery level indicator model

The experimental setup consists of a PIR sensor, accelerometer, and ultrasonic sensor which is interfaced to an NI sbRIO that is acting as a stand-alone system. The stand-alone system is deployed on the AGV that is allowed to navigate across the workplace of the industry. The sensors continuously send the data via wireless router to the user. The Wi-Fi communication is used for this purpose and enables a secured transmission to the user located in the remote location.

The human presence in the path of the vehicle's vicinity is sensed by the PIR sensor by producing a low-voltage signal of 0.312 V to the user. The human emits infrared signal that is detected by the sensor which eventually produces a low voltage at the output. The high logic level 3.45 V indicates the absence of the humans in the path of the AGV. The server code of TCP/IP is deployed into the NI sbRIO, and a client can easily connect to the NI sbRIO by providing the IP address of the router that helps to obtain the exact status of the AGV as shown in Fig. 5.

The presence of an obstacle can be easily notified to the user well in advance with the ultrasonic sensor that is placed in front end of the vehicle. The vehicle stops when the obstacle is detected until the path is cleared. The AGV is able to measure the distance between the vehicle and the obstacle in centimeter (cm). The tilt of the vehicle is indicated to the user using accelerometer that is placed at the center of the vehicle. A predetermined threshold limit 40° is set for the vehicle before it topples. The user can keep a track on the measurement of the degree of the tilt.

The warehouse management system for maintaining the available products will reduce the cost labor and provides an efficient way to keep a track on the stock with minimal effort. With the implementation of warehouse management system, the accuracy of the inventory is increased. The stock in the warehouse can be queried easily with less amount of time. This has made the business logistics to carry out the process very effectively.

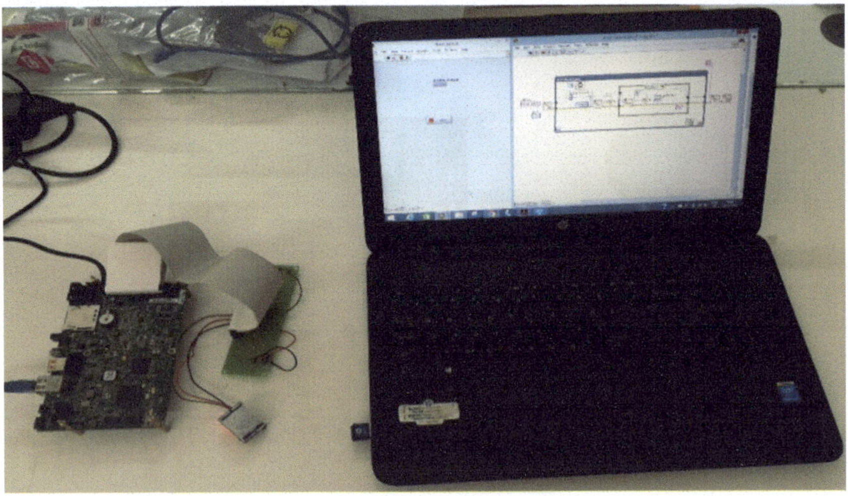

Fig. 5 Wireless data acquisition and communication system of developed AGV

The experiments described in earlier section clearly depict the wireless communication technique that provides a reliable and feasible way of communication between the user and the vehicle. The AGV that operates in a workplace with predefined pattern is easily accessible to the user anywhere from the remote location. The paper proposes the control method of fixed frequency and variable duty cycle that enable to charge the battery in a closed-loop system. The wireless communication system has led to the advancement of the vehicle to monitor more efficiently than the wired network. The authentication procedure for accessing the vehicle is also an added advantage to prevent from intruders unlike other systems.

4 Conclusion

The AGV will autonomously move and keep on updating its status through TCP/IP communication wirelessly via Wi-Fi. Even though the NI sbRIO does not have a built-in wireless module, it is able to support the wireless Wi-Fi router through Ethernet cable. The AGV will automatically move towards the battery charging station whenever it detects the level of the battery low. With the implementation of the battery management system, the charging of the battery is also increased to a higher rate. The elevation of the vehicle due to uneven road surface is detected by the accelerometer. Also the obstacle along the path of the AGV is recognized with the ultrasonic sensors.

The warehouse management system manages the goods present in the warehouse. The quantity of materials can be determined by the user from anywhere in the warehouse. After the goods are being delivered, the user is allowed to update the information such as currently available products in the warehouse. Some products may be no longer exists in the market, and the user is able to delete that item completely from the database. All these functions are carried out by the user using a laptop present in the remote location. The graphical user interface built upon the laptop will help in providing authentication of the user and take control of the vehicle completely. The laptop can be operated anywhere in the industrial warehouse and also provides the destination to the AGV through TCP/IP via Wi-Fi wirelessly.

References

1. Gnanavel Babu A, Jerald J, Noorul Haq A (2010) Simultaneous scheduling of machines and automated guided vehicles in FMS using differential evolution. Int J Prod Res 48(16):4683–4699
2. Michiko W, Masashi F (2001) Intelligent AGV driving toward an autonomous decentralized manufacturing system. Robot Comput Integr Manuf 17(11):57–64

3. Lategahn J, Muller M, Rohrig C (2012) Global localization of automated guided vehicles in wireless networks. In: IEEE international symposium on wireless systems, Offenburg, Germany, Sept 2012
4. Hata T, Ohmae T (2004) Position detection method using induced voltage for battery charge on autonomous electric power supply system for vehicles. In: 8th IEEE international workshop on advanced motion control, pp 187–191
5. Martínez-Barberá H, Herrero-Pérez D (2010) Autonomous navigation of an automated guided vehicle in industrial environments. Robot Comput Integr Autom Guided Veh Ind Environ 26(4):296–311
6. Kongezos VK, Allen CR (2002) Wireless communication between AGVs (autonomous guided vehicle) and the industrial network CAN (controller area network). In: International conference on robotics 8 automation, Washington, DC, May 2002
7. Wiberg PA, Bilstrup U (2011) Wireless technology in industry—applications and user scenarios. In: Proceedings of the IEEE international conference on emerging technologies and factory automation. Oct 2011, pp 123–131
8. Premkumar K, Nigel KGJ (2015) Smart phone based robotic arm control using raspberry pi, android and Wi-Fi. In: IEEE sponsored 2nd international conference on innovations in information embedded and communication systems. Mar 2015, pp 1–3
9. Zhang J, Chunbo CC, Chan A (2014) Wireless power charging method for automated guided vehicle In:. IEEE international electric vehicle conference, Dec 2014, pp 17–19
10. Piyare RK, Singh R (2011) Wireless control of an automated guided vehicle. Proc Int Multi Conf Eng Comput Sci II:16–18

The manufacturer's authorised representative in the EU is Springer Nature Customer Service Centre GmbH, Europaplatz 3, 69115 Heidelberg, Germany. If you have any concerns regarding our products, please contact ProductSafety@springernature.com

Printed and bound by CPI Group (UK) Ltd, Croydon, CR0 4YY
30/09/2026
02225446-0005